Solar Success

Complete guide to home and property systems

Collyn Rivers

SolarBooks.com.au (2020)

Publishing details

Publisher: RV Books, 2 Scotts Rd, Mitchells Island, NSW, 2430. info@rvbooks.com.au

Solar Success

Fourth edition May 2020

ISBN: 978-0-6487945-4-7

National Library of Australia

Cataloguing-in-Publication data

Rivers, Collyn

Solar Success 4th edition

Websites: solarbooks.com.au also rvbooks.com.au/

Feedback: The author appreciates feedback relating to updates, errors and omissions.

Legal: Laws relating to the qualifications required to design and work on electrical installations vary considerably from country to country, (and particularly in Australia). Many require specific licenses for any work above 50 or so volts AC, or 110-volts ripple-free DC. System design too may vary. *Solar Success* strongly advises readers considering any such work to check first with the appropriate regulatory authorities.

Cover photograph:

Kyu Oh, *'Renewable Energy - Solar Modules',* (Getty Images™).

Chapter Listing

Preface	1
Terminology	3
Chapter 1 - Solar Reality - an overview	5
Chapter 2 - Lighting	13
Chapter 3 - Fridges & freezers - Fridge ratings	18
Chapter 4 - Air conditioning	21
Chapter 5 - Washing machines	25
Chapter 6 - Clothes dryers	28
Chapter 7 - Dishwashers	30
Chapter 8 - Power tools	32
Chapter 9 - Phantom loads	34
Chapter 10 - TVs and computers	37
Chapter 11 - Water	39
Chapter 12 - Swimming pools	52
Chapter 13 - Ponds	55
Chapter 14 - Solar	59
Chapter 15 - Solar modules - voltage & current	63
Chapter 16 - Solar regulation	68
Chapter 17 - Batteries	71
Chapter 18 - Battery charging	77
Chapter 19 - Energy monitoring	79
Chapter 20 - Generators	81
Chapter 21 - Alternative power - wind & hydro	85
Chapter 22 - Alternative energy storage - fuel cells	92
Chapter 23 - Inverters	95

Chapter 24 - Energy auditing   100

Chapter 25 - Scaling stand-alone systems   104

Chapter 26 - Meters & measuring   111

Chapter 27 - Installing - legal   115

Chapter 28 - Installing the system   119

Chapter 29 - Constructing a Stand-alone System   147

Chapter 30 - Grid-connect   150

Chapter 31 - Example systems   155

Chapter 32 - Living with solar   165

Chapter 33 - Our solar systems   167

Chapter 34 - Technical Terms Explained   175

Useful sources of information   181

Preface

This book shows how to use our sun's energy to reduce fossil-fuelled: to reduce one's carbon footprint. It differs from much else written about solar in that it stresses the need to reduce that used unnecessarily before thinking about scaling the system.

Reducing one's carbon footprint enables substantial savings in cost.

Achieving the above has complementary and money-saving approaches:

- reducing energy usage by doing things differently
- reducing energy usage by using energy-efficient appliances
- changing the supply of energy to non-fossil fuel by using appropriate technology.

The solar energy required is massively more than enough: we currently use only 1/7000th of that available. It is also affordable – the solar modules needed have decreased in price by 80% since 2012.

China is currently the world leader in the usage of solar and wind power. It had been expected to account for over 40% of all global clean energy by 2022. The 2020 corona-virus outbreak, however, resulted in plans to commission about 140 GW of PV and 75 GW of wind capacity to be delayed. It is (June 2020) expected to reduce commissioned project capacity by at least 20 GW.

Growth in the EU slowed, but Denmark is likely to produce about 70% of all energy from renewable sources (primarily wind) by 2022. If so, it may become the world leader pro-rata population. Ireland is likely to be second. These levels are matched only by South Australia sourcing about 55 per cent of its electricity needs from wind and solar in 2019. It aims for a 'net 100%' by about 2030.

Many off-grid solar generation systems are being installed in the less-developed parts of Asia and also sub-Saharan Africa: output is expected to triple. Much of this growth is expected to come from industrial applications, mini-grids, and solar home systems. One forecast states that this should extend electricity supply to a further 70 million or so people by 2022.

Australia's usage of electrical energy is decreasing slightly (due mainly to many appliances and particularly air-conditioning and TVs becoming increasingly efficient). About 35% of Australia's renewable electricity generation (7.5% of Australia's total) is from hydro-power.

Except for a few tiny islands, Australia has the highest percentage of roof-top solar installations (pro-rata population) in the world. Its solar take-up is also high in rural areas.

The 2020 renewable energy target (of 23.5% of all produced) will almost certainly be achieved. Mainly due to connection delays, transmission losses, and an overloaded grid network investment in large scale renewable energy in Australia plunged in 2019. Investment in Australia's rooftop solar, however, continued to grow. Total installations in 2019 topped 2.2GW (35% more than in 2018). The Northern Territory, Queensland and Victoria have committed to a 40% target for renewable energy sources by 2030.

Growth in the EU slowed, but Denmark is likely to produce about 70% of all energy from renewable sources (primarily wind) by 2022. If so, it may become the world leader pro-rata population. Ireland is likely to be second. These levels are matched only by South Australia sourcing about 55 per cent of its electricity needs from wind and solar in 2019. It aims for a 'net 100%' by about 2030.

Many off-grid solar generation systems are being installed in the less-developed parts of Asia and also sub-Saharan Africa: output is expected to triple. Much of this growth is expected to come from industrial applications, mini-grids, and solar home systems. One forecast states that this should extend electricity supply to a further 70 million or so people by 2022.

Except for a few tiny islands, Australia has the highest percentage of roof-top solar installations (pro-rata population) in the world. Its solar take-up is also high in rural areas. A significant increase is of grid-connect systems in urban areas, but (while not yet economical) there is increasing interest in going off-grid. Properties that had primarily relied on diesel-generated power are increasingly installing solar to provide power throughout much of the year.

Solar Success enables readers to evaluate that offered and to have an educated say in decisions. For those intending to implement stand-alone solar themselves, it assists in ensuring it works first time and every time, and at the lowest possible cost. *Our companion book 'Solar That Really Works'* does likewise for cabins and RVs.

Following the guidelines in this book ensures an economic system that supplies clean and reliable power for years to come. *Solar Success* is intended as a guide to follow. Not a mass of things that have to be remembered. The book's use as a trade text is appreciated but is not the book's primary intent.

Finally - we do walk the walk! We self-designed, built and installed a solar system on our previously-owned 10-acre property in the Kimberley. That system exemplifies that set out in this book. The entire property (home, two offices, a swimming pool and extensive irrigation) ran 100% from solar alone.

Our current home (in Church Point, Sydney) has a 6 kW solar system with Tesla battery storage. We generate two to four times that which we use and currently resell it for 20 cents per kilowatt-hour.

Terminology

Until recently electrical voltages were classified as Extra Low voltage (a voltage that is less than 50-volts AC or 120-volts DC), Low Voltage could deceive (it was 50-1000 volts AC and 120-1500 volts DC). This terminology has changed (internationally) to 'Decisive Voltage Classifications'..

Decisive voltage classification (DVC)	Limits of working voltage		
	a.c. voltage r.m.s.	a.c. voltage peak	d.c. voltage mean
A	≤ 25	≤ 50	> 50
B	≤ 35.4	≤ 71	> 71
C	≤ 60	≤ 120	> 120

NOTE 1: Under fault conditions, DVC-A circuits are permitted to have voltages up to the DVC-B limits for a maximum of 0.2 s.

NOTE 2: If a battery system is either DVC-B or DVC-C it will be treated as an LV installation as defined in AS/NZS 3000.

NOTE 3: The ripple-free voltage value is the d.c. working voltage mean value, where the peak value caused by a ripple voltage of r.m.s. value not greater than 10% of d.c. voltage

Table 1. Summary of decisive voltage classification voltage ranges. From Standards Australia - draft of electrical installations standard DR2 AS/NZS 5139:2019.

Except where necessary for legal and technical reasons, this book still refers to 110-230-volts AC as *grid* or *mains* power.

A further but lesser confusion relates to the term solar 'panel'. To the solar industry, that which almost every buyer calls a solar panel' is a solar 'module', and interconnected solar modules - a solar panel. The industry term for interconnected solar panels is a solar array.

As many solar installers use *Solar Success* as a reference work, this book uses the term solar 'module' in its correct context.

RV Books thanks the Clean Energy Council for assisting, the Australian Bureau of Meteorology for the base solar data from which our maps were prepared, and the Grattan Institute for invaluable background information relating to future solar and electrical industry changes.

Chapter 1

Solar Reality – an overview

Solar is not an infinite resource, but as of 2020, the world's energy needs could be supplied by about 0.007% of that available yearly. Around midday, areas at latitudes like Australia, New Zealand and South Africa have 800 to 1000-watts of solar energy falling on each flat square metre. Currently, affordable solar modules are 14% to 21% efficient. Solar reality is thus about 120-140-watts per square metre.

To use solar energy when no sun is available, we store it in a battery and draw on it later. We can also use the electricity grid as a 'virtual battery', drawing power when we need it, and selling it back when we don't. Some countries pay well for that fed into the grid, but that paid in Australia varies from state to state. By and large (in most states) it currently makes sense to have a solar system that provides a home's 100% electricity needs on a typical sunny winter day.

The Independent Pricing and Regulatory Tribunal New South Wales set a benchmark range for electricity retailers – of (in 2020) 8.5 cents to 10.4 cents per kWh. Tariffs currently being offered by electricity retailers, however, range from 0 cents to 21 cents per kWh.

The highest (as of March 2020) is the Northern Territory's gross feed-in tariff of 23.7 cents a kilowatt-hour.

Queensland's rates vary according to whether you live in southeast or regional Queensland. The so-called 'Voluntary Retailer Contributions' cover current solar feed-in tariffs in Brisbane, the Gold Coast, and the Sunshine Coast (up to Noosa). They're called voluntary because electricity retailers don't have to pay anything for your solar electricity. Fortunately, in most cases, they still do, but it can be as low as 7.84 cents per kW/hour. In practice, many people install solar for other than just monetary reasons.

Figure 1.1: This group of 60 houses in Freiburg, Germany, each produces 6300 kW/h per year. Freiburg is virtually an all-solar town. Most public and commercial vehicles are solar-powered.

How much solar is available?

The amount of sun available depends on locality, cloud cover and time of year. The most you are likely to harvest in a full day is seven times that available for an hour around midday. Three to five times is typical in most temperate areas much of the year, and two to three and half in mid-winter.

Solar input is measured much as is rainfall, i.e. that which falls in a day. By recording solar input this way, peaks and dips average out as the sun (apparently) sweeps the sky, obscured at times by cloud. Irradiation maps show the result in so-called Peak Sun Hours (PSH). Each is much the same as one hour of intense 'noon-time' sun (i.e. 1000-watts per horizontal square metre). Most solar modules are 14%, so 20% efficient equates to about 140-200-watts per square metre.

Meteorological offices have PSH data for virtually any area. Or Google your area plus PSH.

Except for some areas (that have seasonal cloud cover), the change in input from mid-winter to mid-summer is generally smooth.

To estimate probable daily input, multiply the relevant (PSH) by 70% of the marketed wattage of your actual or planned solar array. (Why only 70% is explained later). Now, solar modules are so cheap that whether or not home and business solar are worthwhile depends mainly on your having sufficient sun-facing space.

Solar is marginal at 2 PSH a day: e.g. in a Tasmanian/New Zealand/British mid-winter, but comes into its own from 3.5 PSH onwards. In mid-summer Sydney or Brisbane, there is 6.5 to 7.0 PSH. A 200-watt solar module thus produces about 140-watts times 6.5 hours (910-watt-hours/day).

Figure 2.1. Solar in floating villages on Lake Titicaca (Peru).
Pic: climatechangenews.com

Solar - what runs/what doesn't

Solar works best if you first reduce consumption to the minimum. Doing so requires minor changes – but not necessarily inconvenience.

Water heating is best done by conventional solar water heaters, cooking by gas, and space heating by passive solar and adequate heat insulation. Most everything else can be economically powered from solar, and a great deal of money is saved by using efficient lights and appliances.

Older fridges guzzle energy: this can be halved by using today's more practical examples and installing them correctly. Many water pumps (particularly for swimming pools) are inefficient, but variable speed pumps (Chapter 11) use far less.

Efficient washing machines can be run from solar power and solar hot water. Nevertheless, cold-water washing products do a good job. Front-loaders use less power and water.

Most dishwashers use energy in heating cold water. Some accept heated water, but that's only efficient if its feed is well insulated.

Efficient lighting saves energy, as does turning off unneeded lighting. Incandescent and halogen globes are now banned from sale in most countries.

Fluorescent globes use about 25% of the energy drawn by incandescents. LEDs use even less.

Post-2016 medium-sized TVs are reasonably efficient (those of 2019-2020 are better). The major draw, however, is proportional to the square of the screen's diagonal size.

Laptop computers draw much the same energy as desktop computers. So-called 'gaming computers' draw a great deal of power.

Air conditioning is feasible with solar, but first, reduce that need via ventilation and insulation. Evaporative coolers barely work at all in over 30% humidity. Solar-powered irrigation is feasible, as are domestic swimming pools - but only by using the techniques shown in this book.

Figure 3.1. A village in the tribal area of Melghat, Chickhaldar (India). The Indian government assists solar installation for outside lighting, and each home. Pic: Bspujari © dreamstime.com.

Stand-alone solar systems

Stand-alone systems are used mainly where grid power is unavailable. They are, however, being increasingly considered (instead of grid-connect) as electricity prices increase. Some stand-alone systems are used for pumping water to fill a tank or dam when there's sufficient sunlight, but most serve multiple needs.

Energy is generally needed at all times. It is accordingly stored in batteries and drawn upon when required.

A basic system for a cabin or RV needs only a solar module or two, a solar regulator and a battery. Millions of cheap, reliable and straightforward systems like this are transforming life in developing countries.

For larger cabins and general household or property use, it is more practical to use efficient 230−volt lights and appliances via an inverter. The inverter changes battery voltage to AC mains voltage.

Solar works best where there is little difference between summer and winter sun. Where there is a big difference, it's cheaper to use solar most of the year but supplemented by a diesel generator during mid-winter and extended cloud cover. The generator also runs rarely used heavy loads - such as big arc welders, air compressors or extended water pumping

The balance between solar and battery capacity and other generated power is changing as solar capacity becomes increasingly affordable, battery and fuel prices rise, and rebates fall. It is rarely feasible to rely 100% on solar alone. It *can* be done, but it is hugely expensive.

Grid-connect

Where mains electricity is readily available, 'grid-connect' enables its use as a virtual battery. When more energy is needed than solar-generated, it is drawn from the grid. When there is excess solar, as is typical during summer days, the surplus is fed into the grid and paid for by the electricity supplier.

A drawback of the older grid-connect systems is that if the grid supply fails, your solar power is usually disabled. There are now various solutions, including grid-connect with battery back-up.

What solar modules produce

No commercial solar module in real-life usage produces that which vendors promote. That is only achieved briefly in freak conditions, or a testing laboratory using methods that do not reflect user reality.

Solar modules used in stand-alone systems typically produce 70%–80%, and grid-connect systems 85%–90% of that claimed. A '1.5 kW' system's most probable maximum output is thus 1.25 kW. That this is so is revealed by the solar industry (Figure 4.15), but in terms that mostly only engineers understand. Despite this, buyers are charged and rebates paid for the inflated claims.

Solar access

In Australia at least, there is currently no general legal right to on-going solar access. There are, however,

various ways of protecting your solar access. These include lodging an objection to a proposed development with your local government or creating an express easement or restrictive covenant. Unless there is a prior agreement, however, there is little you can do if a shire or council approves a structure that blocks your system's previous access to the sun.

Figure 4.1. Solar shadowing. Pic: Victoria's Herald Sun.

Intending urban users are advised to consult their local council if there is any possibility that nearby development could restrict future solar access.

Service importance

While efficiency is a pre-requisite, so is repairing stuff that goes wrong. If you live in a major city or large town there are rarely problems. In less populated areas the closest repairer may be hundreds of kilometres away. Do whatever you can to check reliability ratings for the appliances in mind (or at least of their makers).

First all-solar town

Babcock Ranch (Florida) is the USA's first all-solar town. Those living or working in the town can choose to own solar modules or obtain solar from an 880-acre solar field at the Babcock Ranch Solar Energy Center.

That centre has about 700,000 solar modules, and resultant electrical energy is processed and fed into the grid. The town has priority, and any remaining energy is fed into the grid. When the sun sets, stored energy provides uninterrupted power.

Figure 5.1. Babcock Ranch solar power station.
Pic: Jeff Greenberg/Universal Images Group via Getty Images.

Chapter 2

Lighting

Lighting used to consume a fair bit of energy – not just because people had a lot of it, but also because so many people leave lights on when not needed. Also, not helping was the myth that no energy is saved by turning off lights.

The incandescent globes used until recently were primarily small glass-enclosed electric fires that also produce small amounts of light. A few globes are still in use but have long since been banned from sale.

Halogen globes were a less primitive form of an incandescent globe. They produced more light but ran so hot (700° C), they required precaution against fire. Their use was banned (in Australia) in September 2020.

Fluorescent tubes and globes

Fluorescent tubes use 25% of the energy of an incandescent globe of the same light output. They are cheap and effective but bulky. Some people disliked their initially harsh white light, but warm white alternatives have long been available. All conventional fluorescent tubes have a slight flicker, but it only irritates some people. Those tubes with a tri-phosphor coating produce about 15% more light. They are claimed by vendors to last longer.

Compact fluorescents

Compact fluorescents are as efficient as tube fluorescents. They are available in a wide range of shapes, light colour and light output. All are flicker-free. Spotlight and insect repelling versions are also available.

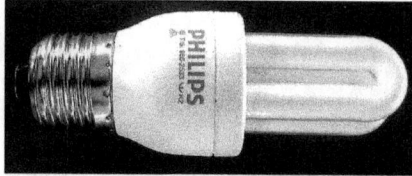

Figure 1.2. A twelve-watt compact fluorescent globe produces much the same amount of light as a 50-watt incandescent: Pic: solarbooks.com.au.

These globes cost more than incandescents but last longer. A few fail within hours, but most last for years. So that faulty units can be replaced under warranty without argument, test them immediately after buying.

Light-emitting diodes

Now almost always known as LEDs, these are rapidly becoming the primary sources of electric light. Once limited in type, today's LEDs produce light in any manner of forms, brightness and 'colour'.

Figure 2.2. This 7-watt LED runs directly from a 12-volt supply. Pic: solarbooks.com.au

Some LEDs run directly from 12-volts via two thin pins that fit into a so-called MR16 base. A few that operate from 9-24-volts plus have a nominally 24-volt supply. They are handy where the existing cable is too light for 12-volt use.

Also available are (MR11 base) LEDs that have similar pins but spaced closer together. The so-called GU 10 types run from 230-volts. There are also 230–volt versions with an Edison screw base.

Light colour

Light 'colour' is based on the concept that heated metal changes colour when heated close to its melting point. White light (4000° to 5000° Kelvin) is primarily associated with fluorescent tubes. It is an excellent choice for kitchens and reading lights, but many perceive it as 'cold'. Many people prefer warm white (2700° to 3100° K) background lighting, bedrooms and outdoor areas.

Lighting levels

Three interrelated units define 'lighting'.

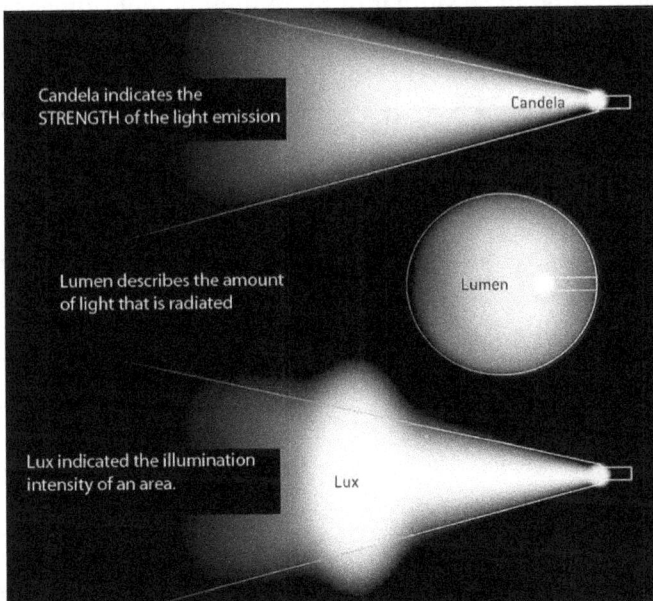

Candela indicates the STRENGTH of the light emission

Candela

Lumen describes the amount of light that is radiated

Lumen

Lux indicated the illumination intensity of an area.

Lux

Figure 3.2. The units most used in lighting. Pic: source unknown.

The basic unit in artificial lighting is the candela. It is a measure of luminous intensity. A lumen is a measure of the total amount of visible light a source produces total area over which the light is spread. Lux is a measure of light falling on a surface.

The units interrelate, i.e., the light level from a light source of (say) 1000 lumens over 1 square metre is 1000 lux. If that 1000 lumen light source is spread over 10 square metres, the average level is 100 lux. In other words, lux takes into account the area illuminated.

Recommended light levels

Most homes need background levels of 50 to 150 lux. General tasks like reading and writing require about 360 lux. High precision work may need 500 to 600 lux.

A lighting level of (say) 100 lux currently (2020) requires about 1.5-watts for each square metre of floor area. A 20 square metre room would thus need about 30-watts (for background lighting).

An LED's efficiency is best assessed by the lumens it produces per watt – it is usually related to its price. A high quality 5-watt LED may thus produce similar or more light output (lumens) than an eBay 'special' of 8-10-watts.

Light type	Luminous efficacy in lumens/watt
Candle	0.3
Incandescent (including halogen)	5-25
Fluorescent tube	50-70 (80 is achieved by the triphosphor type)
Fluorescent globe	45-60
Metal halide	60-115
White LED (and warm white)	100-175
White LED (prototypes)	250 plus

Table 1.2. Light source efficiencies compared. Source rvbooks.com.au

Chapter 3

Fridges & freezers

Air conditioning, conventional electric heating, and old mega TVs apart, fridges and freezers are the most significant energy gobblers in domestic systems. As with air conditioners, they are often perceived as back-to-front heaters that somehow turn electricity into 'cold', but that is not how they work.

A fridge works by having a pump collect heat from its inside (where heat is not wanted) and releasing it where heat does not matter. It does this by circulating a refrigerant gas (that becomes liquid when compressed) through finned tubes inside the space to be cooled. This gas collects the internal heat that is then dispersed externally via finned tubes.

The refrigerant gas is initially compressed, causing it to become a super-heated vapour that is cooled and condensed until it becomes liquid. It then expands into a part gas/part liquid that is pumped through the fridge's cooling tubes - where it absorbs unwanted heat. The refrigerant continues to flow through the system until it dissipates the unwanted heat via the external cooling fins. Instead of fins, some fridges use the side and top outer metal cladding of the fridge to dissipate heat.

Ammonia was used initially for refrigerant but was dropped because it killed people. It was replaced by Freon - that killed the environment instead. A less destructive product then replaced both.

There have been significant improvements in fridge efficiency in recent years. Huge savings in energy consumption can be made by replacing any fridge made before 2000 or so by a post-2012 high-efficiency equivalent.

Fridge ratings

In many countries, refrigerators and freezers have been required to display an energy label and to meet minimum energy efficiency levels. As a result, they are typically 70 per cent more efficient than they were 30 years ago.

Australia's original system allowed for six stars, but by 2009 a few exceeded five stars. A 10-star system was thus introduced from 1 April 2010. The revised system also reduced the number of stars awarded for the same energy consumption.

As a rough guide, a previously-rated five-star fridge is now rated at 3.5 stars. For products that exceed six stars, an additional 'crown' accommodates four extra stars.

Knowing the number of stars assists the buying decision, but indicates only the relative efficiencies of fridges of similar size. A 400 litre, three-star fridge may thus use less energy than a 500 litre, four-star fridge.

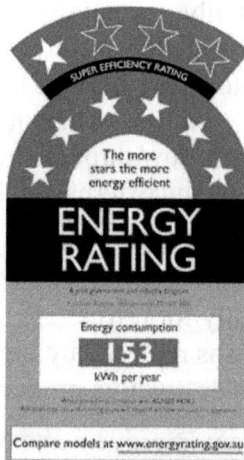

Figure 1.3: The 'Energy consumption' panel below the stars shows how much energy a fridge is likely to use in one year.

Domestic fridges

Generally speaking, the bigger the fridge, the more energy it uses, but not proportionally. A 600-litre fridge is likely to draw less than twice the energy of a 300-litre fridge. Do not buy two small ones rather than one bigger one. Equally, however, do not buy a fridge that is smaller than you need. You may use less electricity, but spend more on fuel as you shop more often. Equally, though, do not buy a fridge that is *much* larger than you need.

Figure 2.3. The Bosch 652 litre
KAN92V130A double-sided fridge.

Many fridges are designed such that they run until the set thermostat temperature is reached. They then cease running until the temperature rises by 1° C to 2° C. At that point, the cycle recommences. The on/off time ratio depends on various factors including set and ambient temperature, what the fridge is cooling at the time, and how often it is opened and closed.

Until recently a typical 500 litre 2000-2002 fridge used about 2.5 kWh/day. There is, however, an increasing move to fridges that run continuously, varying speed to maintain the preset temperature. In 2020 the very best used under 1.0 kWh/day.

Yet more energy can be saved by installing a fridge in a manner that reflects how it works, in particular by thermally separating it from wherever the heat is pumped. Chapter 28 shows how.

By replacing an old fridge with an efficient one of today, you may draw 1.5 kilowatts/hour a day less. Apply the same thinking to everything electrical in the house, and the savings may surprise.

Chapter 4

Air conditioning

An air conditioner works like a fridge with its door open. It collects heat from inside spaces and releases it outside such spaces. It is increasingly feasible to run air conditioning from stand-alone solar, nevertheless, unless you are 100% certain you need air conditioning, consider the following alternatives.

Locate daytime areas and bedrooms in areas that enable natural cooling and heating. Use light coloured roofs and windows to optimise cooling from natural breezes. Minimum interior walls improve airflow. Also assisting is external shading and increased insulation.

It is particularly necessary to stop unwanted heat from entering the cooled area. Add extra or more efficient insulation in the walls and roof, double or triple glazed windows, doors that close without gaps, and external shading.

Air conditioners efficiency varies widely. Reverse-cycle units not only cool but provide efficient heating. In essence, those more efficient are costly, but the savings on the solar capacity to drive them are huge.

Air conditioner star ratings

The initial ratings for air conditioners had six stars. By 2009, however, the most efficient was equivalent to a 10-star rating (had there been such). Current ratings allow for 10 stars, but the old and new ratings are not directly comparable. The new ones are now based on an annual efficiency calculation that includes 'all non-operational energy consumption'. Daikin's US7 2.5 kW unit has seven stars.

Air conditioner types

The two main types of an air conditioner are single unit window/wall systems, and split systems. The latter has one part inside the property and a second part outside. Reverse-cycle air conditioners units provide heating as well as cooling.

Centralised ducted systems were in vogue before global warming became taken seriously. Cooling an entire house is now less common. Single-unit systems are ugly and often clumsy. Old ones are grossly inefficient. Split systems are more efficient and quieter.

Inverter air conditioners

As with older fridges, many air conditioners cycle on and off to attain an averaged set temperature. Again as with fridges, there is, however, a move to running them continuously at whatever speed is required to maintain the desired temperature.

Such air conditioners, known as 'inverter' or 'reverse cycle' units (that can heat as well as cool), cost more to buy, but less to run. If, however, run at their maximum cooling setting (at which their energy rating is determined) they draw more energy than do conventional units. Most air conditioners, however, are run at less than their maximum cooling. At such a part-load operation, inverter units are very efficient.

Reducing air conditioner consumption

Assuming the area is adequately heat-insulated, the energy drawn by air conditioning relates to its set temperature. People may disagree about 'comfortable' temperature as that depends on whatever they are used to, but given time to adjust; 22° C-23° C works well for most. Every 1° C increase increases energy draw by 10% or so. Bear this in mind when scaling the system because a few degrees hotter or colder makes a massive difference to running cost.

Many air conditioners have adjustable louvres. For cooling, these should be turned toward the ceiling. Doing so distributes the cooled air more evenly. For heating, turning the louvres downward directs the heat towards the floor.

Air filters must be installed correctly and cleaned frequently. If obstructed by dirt, cooling performance falls, and energy consumption rises.

Cooling capacity required

The required capacity substantially depends on how well the cooled area is heat-insulated and the outer walls protected against the direct sun. As a guide, 125-watts of cooling capacity per square metre suffices for living areas and 80-watts per square metre for bedrooms. Twenty to thirty per cent less is needed if the area has effective insulation and air sealing.

Reverse-cycle air conditioners

These units heat as well as cool: they utilise heat energy from outside the building. Such operation works because the air that feels cold to us contains heat. The reverse-cycle operation works best in temperate climates where outside temperatures are above 5° C.

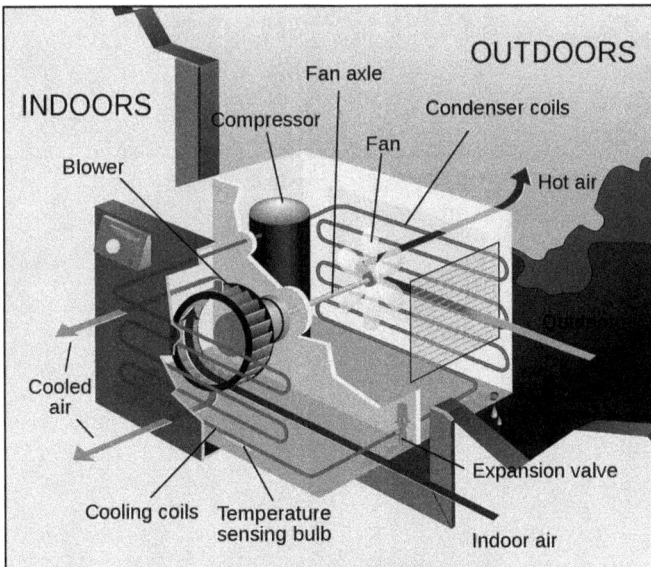

Figure 1.4. How an air conditioner works. Pic: Wikipedia.

Reverse-cycle units have a double panelled label that shows their energy usage for both cooling and heating. A 20 square metre bedroom needs about 1600 cooling watts: readily provided by an efficient unit drawing under 350-watts of electrical energy.

Some reverse-cycle units now produce over 5 kW for every 1 kW of electricity drawn. This type of heating is mainly practicable with solar grid-connect systems that currently have excess capacity.

Ideally, houses need designing with solar heating and cooling as paramount. If that is done well, many need little or no other form of either. Existing houses are often harder to adapt, but as shown in the following chapter, there are also other forms of energy-efficient forms of heating.

Air conditioners (single split systems)	Star rating cooling	Output (kW)	Energy drawn (kW)	Star rating heating	Output (kW)	Power drawn (kW)
Daikin US7FTX225NV1B	7	2.5	0.42	7	3.6	0.62
Mitsubishi SRK20ZJX-S	6	2	0.35	5.5	2.5	0.45
LG K09AWN -NM11	5.5	2	0.46	5.5	3.2	0.59
Kelvinator	5	2.5	0.5	5	2.6	0.52
Samsung AQV09KWAN	5	2.4	0.47	5	3.1	0.6

Table 1.4. Typical leading air conditioners.
Data: energyrating@climatechange.gov.au

Chapter 5

Washing machines

Figure 1.5 Asko W6983 washing machine.
Pic: Asko.

Washing machines vary in their use of water and electricity. The two interrelate: not only must water be used with care, but so also must the electrical energy expended in making that water available.

As a generalisation, front-loading washing machines use less water and less electricity to run than do top-loading machines but cost more to buy. Given the correct detergent, they wash as efficiently as top loaders. They usually take longer per cycle (60 to 90 minutes) but are kinder on clothes.

Top loaders are cheaper to buy, use more water and more electrical energy, and cycle faster (typically 30 to 45 minutes). They are harsher on clothes but usually have a good range of cold water cycles.

Select a unit that can wash using cold water. Washing performance is barely reduced, providing you use the correct detergent. This way, you use 80% to 90% less power per cycle.

Washing machine - energy ratings

The star rating system for washing machines works much as for other appliances. It compares machines of similar capacity. The related data shows consumption for both the cold water cycle and where the machine heats the water, for the hot water cycle also.

As with many electrical appliances, the efficiency of washing machines has improved in the past decade. There are, however, some ancient designs still in production. They are best avoided as the efficiency difference, on warm cycles, between the best front loader, and the worst (slightly bigger) top loader is an extraordinary 11 times. There is not a big difference, however, between machines on their cold water cycles.

Water efficiency ratings

From November 2011 an obligatory additional star rating system applied to washing-machine water usage. Washing machines with a capacity of 5 kg or more had to rate at least three stars. Those with less than 5 kg had to rate at least 2.5 stars.

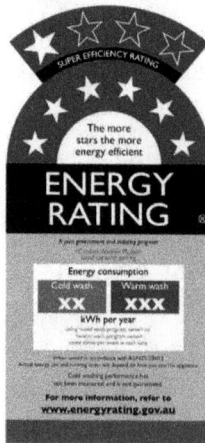

Figure 2.5. Washing machine star rating (2020).

Buying caution

Care needs taking when buying. Washing machine vendors promote their products' water efficiency rather than their energy efficiency. They do so because many products have five stars for water efficiency, but fewer for energy usage.

Rebates

State-controlled rebates apply from time to time for efficient domestic appliances. For a full and up-to-date list of ratings see www.energyrating.gov.au.

Chapter 6
Clothes dryers

It makes little sense to use an electric clothes dryer while the sun is shining. Nor does it make any sense to do so in an electric dryer powered indirectly via solar-generated electricity.

Figure 1.6. Solar - direct: efficient and cheap!
Pic: Hills (Aust).

Dryer types

The most straightforward and most economical are spin dryers that revolve at a higher speed than can washing machines. They do not dry the washing but come close.

The energy drawn by such spin dryers is related to the water removed. In removing *each kilogram* of water, a typical two-star dryer consumes about 1 kWh of electricity. Because of this, the more thoroughly the washing is spun or wrung out before artificially drying, the better.

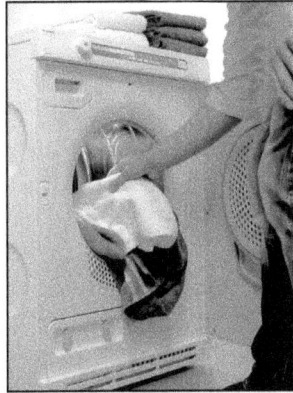

Figure 2.6. Clothes dryer six-star
ASKO-T784HP-01. Pic: ASKO

Significant efforts have been made to remedy this: new technology has resulted in several manufacturers producing six energy star products. One product is claimed to use about 1.5 kilowatt-hours to dry a seven kg load - but costs $2500.

Chapter 7

Dishwashers

Most of the previous chapter's comments about washing machines are true of dishwashers, except that dishwashers are front loading, and that most draw cold water that they heat for use. A cold water connection is not that wasteful as dishwashers only heat water for part of the cycle. Some do accept hot water so, if you have a solar water heater, it makes sense to use it for this.

Figure 1.7. SMEG stainless steel dishwasher (DWA157X) can be built-in or free-standing. Pic: Smeg.

Energy ratings

With one major exception, the dishwasher star system and energy ratings are almost identical to those relating to washing machines. The exception is that dishwashers are rated against a specially made and calibrated reference machine, a Miele G590 dishwasher. It is from this that the official dishwasher Washing Index and Drying Index is derived. The tested dishwasher's performance must exceed that of the reference machine.

With most appliances, it makes no sense to buy one bigger than you need, but dishwashers are an exception. Most that handle 10 or more table settings use less energy and water than those with fewer place settings: less absolutely – not pro-rata their size.

No matter the brand dishwasher you have, you can save energy and water by totally filling it before running it, choosing an economy cycle, and avoiding heat drying options.

Brand & model	Star rating	Place settings	Energy usage
Siemens SN26E281AU	4	15	670-watts
Bosch SM150E45AU	4	14	616-watts
SMEG DWAFI152T	4	15	670-watts
Baumatic BDW65W	3.5	14	723-watts
Fisher & Paykel DW60CCW1	3.5	14	745-watts
Venini VDW71W	3.5	14	724-watts
Hotpoint HPDW06BISS	2	6	523-watts
Classique CL45DSS	1.5	8	805-watts
Andi AF456EX	1.5	8	805-watts

Table 1.7. Many large dishwashing machines use less energy pro-rata than smaller ones.

Saving energy

Most dishwashers and liquids are intended to cope with scraped but unrinsed dishes. Don't waste hot water by prior rinsing. Run the machine only when it is full and either use the economy drying cycle or, better still, let the contents dry naturally.

Chapter 8

Power tools

Most hand-held power tools have motors that are efficient while running, but draw several times their running current for a second or two while starting. A half-horsepower (375-watts) drill, for example, is likely to draw 450-watts while running, but 1000-watts or so while starting.

Figure 1.8. Some power tools, such as this Makita 230 mm angle grinder have a soft start. Pic: Makita.

Older (and large) transformer type inverters can usually start any hand-held power tool they are capable of running. The later transformer-less switch-mode inverters have next to no such excess capacity. The list below shows approximate *running* currents. You will, however, need an inverter capable of supplying three times the running current of that tool.

Big motors, such as those that drive cut-off wheels, may draw a momentary 6000 to 9000-watts, but some have a 'soft start' feature that limits starting current. Those that do usually have this made clear in their promotional literature.

Equipment that draws massive starting current, or high peak loads (such as welders) start more readily from an already-running diesel rather than from a petrol-driven generator. The explanation is that the former invariably have heavier flywheels whose momentum assists overcoming the load's high starting draw, and also changes in load.

It is feasible to run such equipment from solar and is often done with large systems used on outback properties. By and large, though, trades-people using heavy equipment use portable generators.

Item	Typical running power draw (in watts)
Air compressor 1/2 HP	750-1000 (but starting surge may exceed 5000)
Circular saw (100 mm)	1400-1500
Circular saw (250 mm)	1750-2000
Drills (hand-held)	350-650
Drill press (bench/pedestal)	500-1000
Garden blower	2200-2500
Guns – heat	1500
Grinders (100 mm)	600
Grinders (200 mm)	900-1000
Jigsaw	500-600
Sander/polisher (orbital)	150-250
Sander (sheet)	175-250

Table 1.8. The typical power draws of tools and devices.

Chapter 9

Phantom loads

All mains voltage devices draw power if left on their 'remote' power setting. So do many that have an external power supply or voltage adaptor, those with continuous displays and those that charge batteries. A typical home has thirty or more, and all need switching off at their wall outlets.

EU and USA legislation now limits these so-called 'phantom loads to 1-watt, and some to 0.5-watts. Much imported product, however, still exceeds these amounts by many times.

Each phantom load that draws one watt consumes almost 9 kWh a year. That typical thirty of them consumes about 270 kilowatt-hours a year. According to Australia's Department of Industry, Innovation and Science, that is 5.9% of Australia's total residential electricity use.

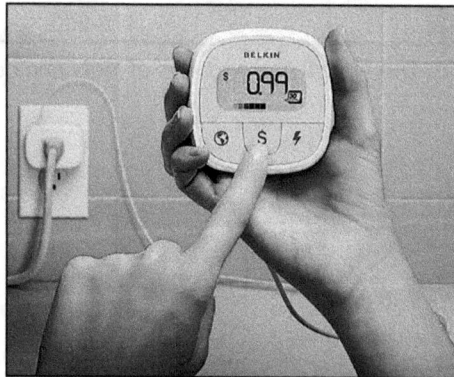

Figure 1.9. The Belkin F7C005au energy meter – basic but accurate. Pic. Belkin.

Example phantom loads

A well-known 'instant' coffee machine draws 180-kilowatt-hours/year (about half the draw of a 350-litre fridge) for trivial convenience. Such usage is justifiable in office environments but is pointless in homes.

Many TVs are left on all day and night (with the sound turned down while not being watched). Computers and printers continue to draw power when switched off. So do washing machines and dishwashers.

Mains-powered door chimes may draw 350 kW hours/year yet be used just a few times a week and for a few seconds each time.

A significant concern is adaptor plugs that feed 12–volt lights and appliances. The larger ones draw power whether the appliance is on or off.

There's nothing worse than receiving a scary energy bill. When it arrives, you may run through in your head everything you did that month, yet find nothing that could account for how high your bill is. But did you take into account phantom and standby energy draw?

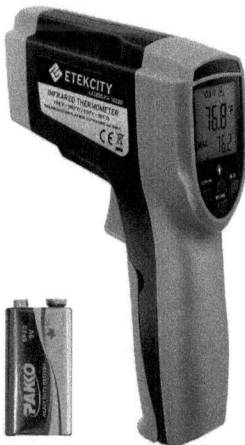

Figure 2.9. This Etekcity 1022D infra-red thermometer measures from 50-550 degrees C (58-1002 F). Pic: Etekcity.

Remedying phantom loads

Check by plugging the (turned off) appliance via an energy meter into the wall socket and turn on the power. If a phantom load exists, the meter shows power being drawn. Also, check the draw of equipment turned on but not in use (e.g. TVs left on with the sound turned down).

Device	Watts	Daily draw (watt-hours)
Video game (hand-held)	1	24
Clock radio	1.5-1.75	36-42
Cordless telephone	2-2.5	48-60
Answering machine	2.5-3.5	60-84
Microwave oven	3-3.5	72-84
Typical TV	4-5	96-120
VCR/DVD	5-6	120-144
Dishwasher	6.5-7.5	156-180
Cable TV box	11-12.5	264-300
Security system	17.5-20	420-480
Coffee machines (some)	20-25	480-600

Table 1.9. The typical draw of phantom electrical devices in 2020.

Caution is needed when buying an energy meter, as many are inaccurate at low power levels. A quick way to spot phantom loads is with a hand-held infrared thermometer (Figure 2.9). You use it to check if a device is warmer than its surroundings.

Chapter 10

TVs and computers

There have been significant and ongoing improvements in TV technology since the first edition of this book was published in 2008. Then, the average energy draw of a typical 80 cm TV was about 145-watts. In 2018 that TV drew only 60-watts and now (2020) about 50-watts.

Sales of LED TVs are second only to the usually cheaper LCDs (Liquid Crystal Display), but plasma TVs are now rare.

TV energy draw

As with any video screen, energy draw increases in proportion to the square of the screen's diagonal size, but there are substantial variations from brand to brand.

Energy usage is also a function of screen brightness, but TVs have provision for reducing this. Retailers turn the brightness to the maximum on display units. It is worth reducing the brightness, and also of the viewing area's lighting.

The previous 10-star rating system changed in April 2013: a previously four-star rated TV was thenceforth rated as one star. Energy requirement regulations, too, were tightened. Product imported before that date could be sold without restrictions, but from October 2013 all had to meet the four-star rating (corresponding to one star on the new scale). There are now many 8 and 9-star products. See http://reg.energyrating.gov.au/comparator/product_types/

Care is needed when buying as some of the more costly TVs are priced well above those that are far more efficient.

Computers

Some US-made computers are (US) 'Energy Star' rated, but there is no Australian equivalent. The screens have similar energy usage patterns as TVs.

Significant energy reductions can be made, particularly with desktop computers, by setting them up in an efficient power management mode. This mode is usually accessed via the operating system. In some cases, energy usage can be slashed by up to 80%.

Plug-in power supplies for laptop and notebook computers draw power when the computer is turned off: switch them off at the wall. Unplug your power supply after the notebook battery is charged, or use a power board with on/off switches.

Figure 1.10. Panasonic's 125 cm (50 inches) TH-L55ET60A TV draws about 60-watts. Pic: Panasonic.

Some operating systems save active programs and files before shutting off and restore them when the computer is turned on - encouraging users to close their computers when not in use. If yours must remain on in the evening for file backup or other purposes, turn off the monitor.

Most laptop computers and notebook computers are more efficient than their desktop equivalents. Their semi-battery operation necessitates that.

Chapter 11

Water

For non-potable use, rainwater needs only mechanical screening. For other uses, rainwater needs filtering to remove airborne pollutants, those from roofing and roof flashing materials and animal and bird droppings. Screens are essential to prevent animals and birds entering the tank – and becoming trapped.

The main requirements include a first flush diverter at the tank inlet to prevent the initial contaminant laden water from the roof entering the tank when it rains. That device and catchment area needs cleaning a few times each year. A coarse mechanical screen filter is needed to catch twigs, etc. on the input into the tank. It is also necessary to ensure mosquitoes cannot enter and breed in the tank.

The primary health risk is from pathogens, such as cryptosporidium and giardia. A primary 10-micron filter (and secondary 1.0-micron filter for drinking and cooking water) are typical in outback use. Airborne pollution is less there, but many small things crawl and fly. Local councils, state health authorities or rainwater tank suppliers can advise regarding this.

There is no official standard for filters and housings, but a de facto 'standard' enables some interchangeability. Filters that fit only their maker's housings cost more.

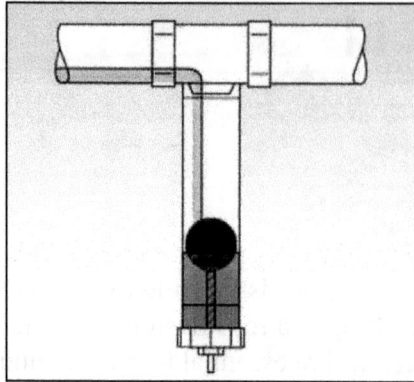

Figure 1.11. First-flush water cleaner – simple but effective. Pic: Water Diverters.

Rainwater tanks

Above-ground rainwater tanks are the cheapest. Slimline and wall line tanks fit narrow spaces but are more costly per litre.

Underground tanks save space and free up catchment area, but, as excavation is needed, installation is costly. For people that rent or likely to move, a cost-effective approach is to use wheeled garbage bins joined by flexible Polypipe.

Locate the main tank as high as possible, and collect the rainwater in a smaller tank that holds several hours' downpour, and then pumped up later via solar power. If the main tank is high enough, the return feed can use gravity alone. Using 40-50 mm piping reduces friction loss.

Use Blue-Line hose and fittings for water connections. The product is made in both metric and imperial sizes. A few sizes are almost identical but not totally. In Australia, metric fittings are usually easier to find.

Rainwater legislation

Rainwater usage regulations can be obtained from the local council and state health authorities. Tank water rebates are currently available from state and territory governments and, in some areas, from local councils. If rainwater is to be the only source, the tank needs to be 50,000-100,000 litres. For garden watering only, a typical urban site needs a 2000-4000 litre tank.

If a rainwater *and* mains supply are used, the mains water system must be isolated from the rainwater system by a valve mechanism or tap. Your local council or state health department can advise.

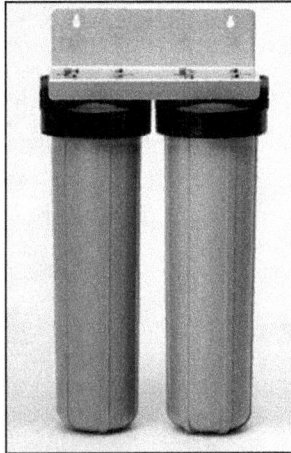

Figure 2.11. Most potable water systems have a fine mechanical screen plus a 10-micron and a 1.0-micron filter. Pic: Big Blue Water Filters Inc.

Pumping water

Where the location permits, an effective technique is to use solar energy to pump stream, bore or rainwater to an elevated storage tank. This tank needs to be at least 20 metres above main outlets, such that gravity supplies the return pressure. Doing this works particularly well where the uplift pump is programmed to run only when surplus solar energy is available and overridden to run at other times if required. Many solar regulators and energy monitors have inbuilt programmable switching that can do this, via a suitably rated contactor.

The suggested 20 or so metres provides a useful working pressure of 200 kPa (29 psi). That pressure is sufficient for domestic taps and is acceptable for most washing machines and dishwashers. Timed irrigation valves work down to 50 kPa (7.2 psi) but may act erratically, or not at all, below that.

In areas, where gravity feed is not possible (or to boost gravity feed), use domestic pumps to supply water for household use. Also, when authorities permit, for watering gardens.

Cabins and RVs commonly use small pumps that have pulsating diaphragm valves driven by 12/24-volt DC motors. Pumps such as these are reasonably reliable but dislike not being regularly used. They may need overhauling yearly. None is intended nor suitable for continuous use. They are also noisy.

For larger dwellings and properties, 230–volt pumps are available from irrigation and other suppliers. Most use a centrifugal impellor driven by a standard induction motor. They are quieter and more reliable than diaphragm pumps but, in sizes below 150-watts, their motors are less energy efficient.

Pump energy draw

Australia does not currently require pump makers to meet specific energy requirements (except for three-phase units above 0.73 kW). Pump makers can, however, choose to submit pumps for testing and receive a star rating. This situation may change, so it is advisable to check the current situation: energyrating.gov.au.

Pressure systems

The simplest way of supplying water via an electric pump is to have a pump switch located close to each tap. Doing so is cheap and ultra-simple and provides water at more or less constant pressure. Using it, however, may necessitate using both hands (to adjust temperature). Moreover, installation requires electric cabling and a switch for each tap. Doing this is feasible for 12/24-volt pumps, but is not feasible, safe, or legal for 230-volt pumps.

Water pressure sensing control is more convenient and efficient. Open a tap, pressure drops, and the pump is switched on: close the tap, pressure rises, and the pump is switched off again. Such a system is simple to install but has several drawbacks in its basic form.

The main issue is that the pump kicks in almost instantly that water is drawn. Then, unless the tap is fully open, water pressure rebuilds rapidly until the pump stops. Such pumps may cycle every second or two, causing water pressure to rise and fall constantly. Washing machines and dishwashers do not like this, nor do people having showers, not least because the water temperature may vary as pressure varies.

On/off control causes a pump to start and stop thousands of times a day. On/off operation shortens pump life and, because they momentarily draw twice or more their running current each time they start, energy usage is high. The pump tends also cycle on and off because even minor temperature variations cause pressure changes in flexible hoses. A slightly leaking tap does likewise.

Most pumps are designed for specific flow rate, pressure and head. Some are designed to supply large volumes of water at low pressure, others to pump less water at high pressure. Specific design works well for reasonably constant loads, but not where loads vary. For efficient domestic and similar use, variable speed pumps, or the pressure tank system (below) are acceptable compromises.

Constant pressure pumps

Constant pressure pumps run at full capacity even if a tap is only slightly dripping. Non-needed water is pumped around an internal bypass in the body of the pump. The concept is like driving an automatic transmission car with the accelerator pedal held to the floor and controlling speed via the brake pedal. They work well, but are far from energy-efficient and thus not recommended for solar.

Variable speed pumps

These pumps vary their motor speed according to the flow or pressure required. There are two main types. One varies the frequency of the power to an otherwise standard pump motor intended to run at 3000 rpm (revolutions per minute) from 50 Hz. A costlier but more efficient approach uses a brushless DC motor and controller resulting in energy reductions of 70%–80%. (Personal experience confirmed this is realistic.)

Brushless DC motor pumps work well for circulating water in swimming pools - where the power needed for occasional cleaning is several times higher than needed for ongoing circulation. (see Chapter 11).

Pressure tanks

A pressure tank enables water to be supplied primarily by air pressure. Given a large enough tank, this reduces pump cycling – from many hundreds of times a day – to only once or twice.

The tank contains a balloon that is inflated to just below the water pump switch's cut-in pressure. When the pump is initially turned on, it forces water into the tank, partially filling it and compressing the balloon. Pressure builds up until cut off by a pressure switch. At this stage, the tank is typically half full of water, and half full of the now compressed balloon.

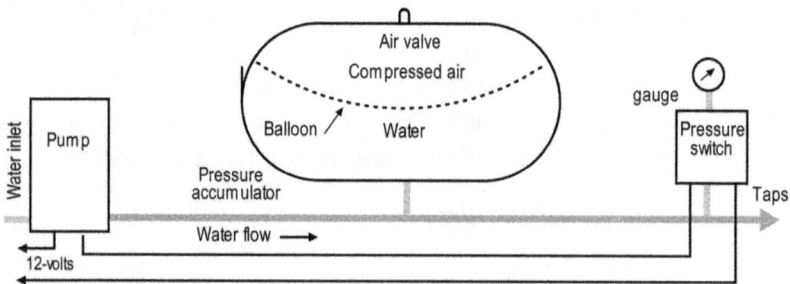

Figure 3.11. Basic pressure system. Tank and pressure switch is best located within 20 or so metres of the pump. Pic: solar books.

When a tap opens, the balloon provides water pressure until the balloon's pressure reaches the pump's cut-in pressure. The pump then starts up, and the cycle recommences.

This method works best with large pressure tanks. One of 500 litres supplies about 260 litres of water, using air pressure alone. Using one of this size necessitates the pump to run only once or twice a day, typically for four or five minutes each time. The pressure drops very slowly, and the change is rarely detected by users. The pump is chosen to match the requirements of the pressure tank so runs at maximum efficiency.

Figure 4.11. These 1.8-metre high 500-litre pressure tanks were buried for cyclone protection. Pic: solar books.com.

Our previously owned 10-acre property north of Broome initially had a constant pressure 70-watt pump that drew over 1100-watts a day. It was replaced by a 450-watt pump and two 500-litre fibre-glass tanks. Tap pressure dropped slowly during the day but was barely noticeable. Energy draw was reduced to a little over 100-watts/day with no detectable user difference.

Pressure tank operation works well with most washing machines and dishwashers. They tend to be affected by rapid pressure variation but not by the slow change of a pressure tank system.

Figure 5.11. Aquatec pressure tank and pump.
Pic: Aquatec.

Irrigation

Property irrigation systems often rely on bore water. The usual practice is to use an on-line bore pump, but when (not if) the pump packs up, or the bore requires cleaning, irrigation stops too. The following approach was designed to provide a week or so of ongoing water at all times. It can be used with any readily available bore pump that matches the bore's depth and water flow required.

A solar-powered bore pump, controlled by a timer, begins pumping up water when the main batteries are fully charged (typically before midday) into a nearby holding tank. An upper float switch cuts off the pump when the tank is full.

A secondary float switch starts the bore pump at any time if the tank level drops below 60% full. A second pump provides the irrigation feed and supplies external taps, etc. This dual-pump system uses more power but ensures ample water is available in bushfire prone areas. The primary need is for large amounts of water to be quickly withdrawable from tanks. Most bush brigades throw a loose hose into the tank, but some have facilities to draw from a typically

1.5-inch or 2.0-inch outlet. Check with your local fire authority for size and thread requirements.

Energy draw is substantially reduced by minimising pumping losses. Select large-diameter Polypipe for main runs, teeing off thinner piping for short runs to drips, etc. Working pressure can be as low as 70 kPa (10 psi) yet still provides adequate drip flow and minimises pipe leaks and failures.

Drip-feed systems work well, but filtering is essential. They need regular checking for blocked jets and piping dislodged by animals etc. Problems can be more readily located by having individually valved irrigation sections – each with its pressure gauge. Leakage shows as abnormally low pressure. Blocked jets show as abnormally high pressure.

Valve control

A problem with using automated irrigation valves powered by solar is that some valves need electrical power to keep them turned off. They are opened by power being temporarily removed and closed by power being restored. These valves are fail-safe in that they water if power is *interrupted*, i.e. they use electrical energy almost continuously except when watering. Their energy draw becomes serious if many are in use.

One solution for the above issue is battery-powered 'one-shot' valve controllers. These generate a mechanical pulse that turns on an associated water valve. At the end of a preset time, a second pulse turns the valve off again. The valves are programmable for days and times. They are long-lasting and reliable. Most need a minimum water pressure of 10 psi (69 kPa).

If the irrigation system's pump can only just supply sufficient water, a pressure tank is of no value. It is, however, if the irrigation pump has a more abundant flow than otherwise needed. In practice, a larger pump plus a pressure tank also provides pressure for other uses while irrigating.

Bore pumps

Bore pumps have a hard life. The 12 or 24-volt diaphragm versions are practical and efficient (they run directly from a solar module, i.e. no regulator is needed) but tend to need servicing yearly. For all but basic cabins, it is better to use a 230-volt bore pump driven via an inverter. These pumps are designed for specific flows and pressure heads. It is essential to select the right one. Unless sure of your ability to do this, seek advice.

Variable-speed bore pumps work well, but basic ones are cheaper, more readily available and more reliable. They can also be fixed without needing specialised service. Bore pump life varies: you are doing well in many areas if you get over four years of use.

Figure 6.11. Honda 5.5 hp petrol-fuelled fire pump. Invaluable for fire-prone properties. Pic: Honda.

It is better not to rely on electrical power being available for fire pumps. The smaller petrol-fuelled ones are perfect for putting out spot fires and light enough to carry.

Pumping losses – the pipe size and power required

When energy was cheap, and people less energy conscious, little regard was paid to the energy needed to force water through pipes. It only *appears* to flow smoothly. In reality, water resists being pumped.

That resistance is expressed as the equivalent height to which it can be pumped compared with pushing it through 100 horizontal metres. It is aptly called 'head loss' and expressed in metres.

The power needed to pump water through 100 metres of 20 mm polythene pipe at 12 litres/minute is the same as that required to raise it 11 metres. This effect is called (an equivalent) head loss of 11 metres. For 25 mm pipe, the equivalent head loss is 3.7 metres, for 32 mm it is 1.1 metres, for 40 mm it is 0.4 metre and for 50 mm, only 0.1 metres. Each tight bend adds about 5%. Sweeping bends are preferable.

The total head of water must include the end pressure required. For non-mains pressure water, this is likely to be 150 kPa (21.7 psi) to 300 kPa (43.5 psi): 150 kPa is equivalent to 15 metres of head loss.

Flow rate and friction loss – PN 12.5 high-density Polypipe – per 100 metres						
litres/min	20 mm	25 mm	32 mm	40 mm	50 mm	60 mm
12	10.9	3.7	1.1	0.4	0.1	-
24		13.4	3.9	1.3	0.4	0.1
36			8.3	2.8	0.9	0.3
48			14.2	4.8	1.6	0.5
60				7.2	2.4	0.8
72				10.1	3.3	1.1
84				13.5	4.4	1.5

Table 1.11. This data shows the pressure loss if too small pipes are used: it is based on data supplied by Grundfos.

The total loss is the sum of friction loss, bend losses, the height through which the water is raised, and the pressure required at the taps. If pumping over distance, it is worth spending more on bigger pipes and fittings. Doing so enables the use of smaller, cheaper pumps that also draw less energy.

Buyers often (wrongly) assume that if a pump is rated at 750-watts, it consumes 750-watts of electricity. Pumps are rated in terms of work done (and energy transfer is never 100% efficient).

The work pumps perform is usually shown as P_2. The energy drawn in such pumping is shown as P_1. Both are expressed in watts, but often only the former (work that was done) is shown. A pump rated at 750-watts is thus able to do 750-watts of work. That amount of work is equivalent to raising 225 litres of water 20 metres in one minute (or 1 HP in our previous terminology). As can be seen from Table 1.13, the most significant loss is usually due to pipe friction.

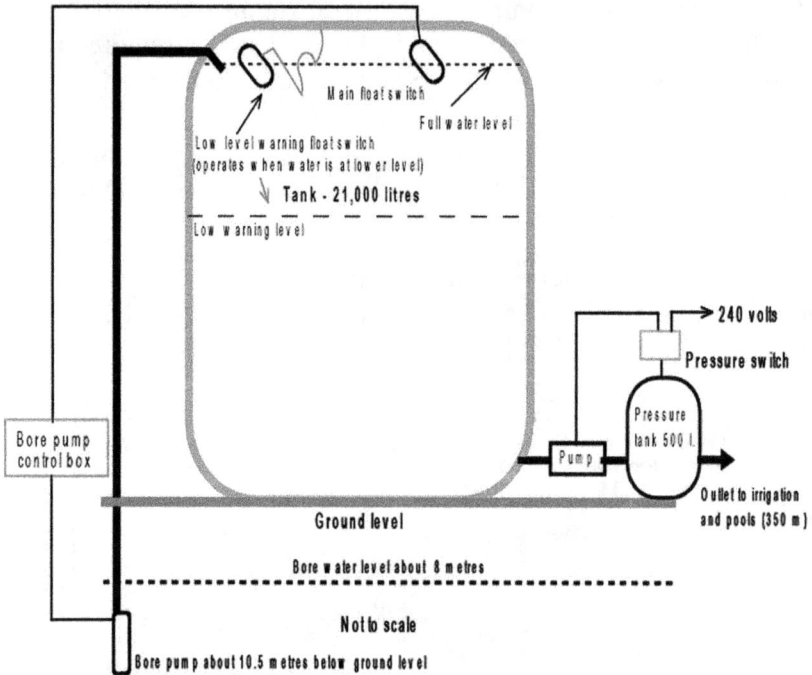

Figure 7.11. Our (previous) twin pump irrigation system.

The bore pump (in Figure 7.11) started pumping about noon. It was turned off when the water level triggered the upper float switch. A second float switch reactivated the bore pump if the water level dropped below 60%. The system was used for irrigation and a 30,000-litre swimming pool. It ran on 12-volts DC. Pic: solar-books.com.au

Calculating pump pressure

To estimate pumping pressure required use the formula: – Hd – Hs + Hf + Pr = P

Where: Hd is the height (in metres) between the pump and highest point. Hs (in equivalent height in metres) is any pressure already available. Hf is the total friction loss for the flow rate. Pr is the pressure required at the tap/s. For example, Hd is 20 metres, Hs is

0, Hf (at 48 litres/minute through 50 mm pipe) is 1.6, and Pr is 150 kPa. Then 20 + 1.6 + 150 = 171.6 kPa (about 24.5 psi).

For most pumps, pressure varies as rpm², the power drawn varies as rpm³. One horsepower is the equivalent of 750-watts, 1 psi is 6.9 kPa (kilopascals).

Chapter 12

Swimming pools

Domestic swimming pools need water recirculated and filtered at least every 24 hours. To do this, most have 230–volt induction motor pumps. They can be driven by solar, but doing so is inefficient and costly.

When initially planning our (Broome) swimming pool, all suppliers insisted it is done via 3-5 kW of solar modules, or by doubling our primary system. Quotes averaged $60,000. We knew there were better ways of doing this. No supplier would consider any, so we designed and installed it all ourselves.

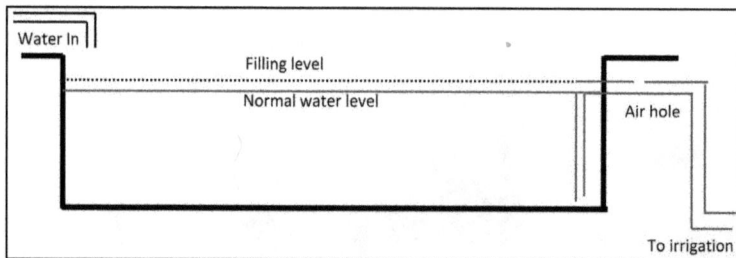

Figure 1.12. A timer valve (not shown) allows irrigation water to flow for about 30 minutes each morning. As the water level rises above the bottom of the U-shaped pipe at the right, it flows down and into the irrigation system. The air bleed hole is essential, or the pool siphons dry!
Pic: solarbooks.com.au.

Calculating the capacity needed to recirculate 35,000 litres of water showed this to be feasible via a 48–volt DC brushless motor and an MPPT controller. The pump's 450-watt draw was handled with ease by four dedicated 120-watt Kyocera modules on cyclone-proof mounts via Broome's daily 8 to 10 hours of intense sun. There was no point in pool water at night – so no batteries were used.

Pumping losses were minimised by using oversize filters and piping. The cost, including concrete and steel for the solar mountings and all the electrics, came to less than $8000.

Keeping water clean

Pool water is usually sterilised by adding chloride or passing lots of amps through saltwater to produce it. The former did not appeal, and the latter required too much power. The solution was diverting the existing irrigation system's daily 2000 litres to flow via the pool controlled by a valve that opened 30 minutes each day.

We ran a 50 mm pipe from close to the bottom of the pool's far end and extended it vertically to the pool's normal water level. The pipe then ran horizontally for about 30 centimetres, and then down into the existing irrigation system. When the valve opened, the water rose to the horizontal section and down and into the irrigation system. The horizontal pipe has an air breather hole opening in its upper surface to allow air to enter. Without it, once the water starts flowing, it continues to siphon until the pool runs dry.

Figure 2.12. Lorentz Badu 48-volts DC unit pumps 35,000 litres a day using four dedicated 120-watt solar modules. The apparent rust is the Kimberley's Pindan sand staining everything it touches. Pic: solarbooks.com.au

Recent trends

To cope with pool cleaning the pump customarily used in domestic pools is about four times larger than needed for circulation. In an attempt to save energy, some installations have a large pump for cleaning and a second smaller one for circulation.

A better solution is one of the high-efficiency pumps that are variable speed (230-volts) versions of the 48-volt Badu pump. The

Badu EcoM3-Speed brushless DC motor swimming pool pump has three main switched ranges, each providing variable speed control. The overall range is from 1000 to 3450 rpm and suitable for pools from 25,000-85,000 litres.

Figure 3.12. The Broome pool. The four 120-watt modules can just be seen low down at the far (northern) end of the pool. The water nearby is a tidal lagoon - further back is the Indian Ocean. The auto water level pipe is at pool right. Pic: solarbooks.com.au.

Davey's PowerMaster ECO is an eight-star efficient pool pump. It has a three-speed 'intelligent' control for water circulation and filtration, automatic pool cleaning, and a high flow option. Davey claims it uses up to 70% less energy than traditional pool pumps.

Various makers claim reduced operating costs, lower energy usage, and reduced noise. While some later systems are more sophisticated, none appears to compete price-wise with that low cost, all solar system in Broome. It is still working after 20 years.

Chapter 13

Ponds

Outdoor ponds need their total volume of water to be recirculated every two hours or so. Position the water inlets and outlets to ensure water is flowing throughout the pool.

Ponds work best if there's a right balance of fish and aquatic plants to prevent algae build-up. The water can, however, also be kept bright by additives or by an ultra-violet steriliser (as described below). Before installing a steriliser, firstly experiment with additives.

Figure 1.13. This self-made 5000-litre fish pond on our previous solar-powered Broome property had water circulated by a submerged pump under the rock (bottom left), discharging from a hollow log about 0.5 metres above water level at the far end. The 12-volts (25-watts) AC pump was run from the central 230-volts 3.4 kW inverter. Pic: solarbooks.com.au.

Ultra-violet sterilisers

Inline UV sterilisers have a small internal UV light. The light is sized such that it draws about 1-watt per 100 litres/hour of water flow.

Blanketweed (spirogyra) can be controlled by an additional unit that, via a cable wrapped around the pipe, disrupts the calcium ions in the water. The units draw a nominal 1.5-watts.

Pumps - general

For ponds up to 5000 litres or so, submersible pumps are a good choice. They are silent and require next to no installation. The better examples are reasonably energy-efficient. Many have a magnetic drive from the motor to the impeller - that increases reliability.

Figure 2.13. Vitronic 18-watt UV filter copes with 2200 litres/hour flow. Pic: Oase.

Given the right pump, power needs are minimal but, as with water pumps generally, pond pumps must match the application. The usually specified pipe is about 12.5 mm, but doubling that reduces losses (and energy used) by five times. Ideally, keep pumping height to a metre or so, or the pumped volume falls off rapidly.

Figure 3.13. Fish Mate pond pumps can pump up to 20,000 litres/hour. Pic: Fish Mate.

You'll need some sort of inlet filter to protect the smaller fish from being blended by the pump's spinning impellor.

Solar pumps

While 12/24-volt DC pumps that run *directly* from solar are cheap and straightforward, most have a lifespan of 1000 hours running (some less). A better approach is a fully submersible pump. This type of pump has a 12 or 24-volts AC motor powered via a small safety isolating transformer. This motor can be run via grid-connect or the central inverter for stand-alone solar.

You may legally self-install pumps sold with a plug-in power supply and specifically made and approved for external use. If in doubt, consult a licensed electrician.

Ponds requiring over 5000 litres/hour usually have an external pump. Their location requires care as the pumps tend to be noisy.

Figure 4.13. PumpMate PM submersible (24-volts AC)
1500PLV pumps 1450 litre/hour. Pic: PumpMate.

Model	Watts	Head (m)	0.5m ltr/h	1m ltr/h	1.5m ltr/h	2m ltr/h	2.5m ltr/h	3m ltr/h	3.5m ltr/h	4m ltr/h	5m ltr/h
Aqua-max2000	32	2	1500	990	510						
Aqua-max3500E	45	2.2	2700	1980	1140	300					
Aqua-max5500E	60	2.8	4320	3300	2100	1140	540				
Aqua-max8500E	80	3.2	6900	5520	4320	3000	1740	540			
Aqua-max12000ED	130	5.2	10,800	9600	8700	7500	6300	5100	3900	2700	540
Aqua-max16000E/D	170	5.6	14400	13200	11700	10200	9000	7500	6000	46800	1800
Neptun300	4.5	0.5									
Neptun600	7.0	1.2	360	180							
Neptun1500	18	1.8	1080	750							
Neptun2000	25	2	1260	1140							
Neptun3000E	40	3.2	2640	2250	1800	1260	960				
Neptun4000E	50	3.4	3600	3150	2700	2100	1500	750			
Neptun6000	110	5	5550	5040	4500	3900	3300	2700	2100	1350	
Neptun9000	195	6	8520	8040	7500	6900	6300	5640	4860	4050	2250
Neptun12000	270	7	11400	10800	10080	9432	8700	7950	7200	6300	4500
Profinaut21	470	9.6	21000	20400	19800	19200	18600	17400	16200	15000	14400
Profinaut27	645	10	25800	25200	24600	24000	22200	21600	21000	20400	18000

Table 1.13. Energy draw and flow rates for typical high-quality pumps. Note how flow rate drops off as the pressure head rises. By limiting the height the water is pumped, and by using piping of twice the diameter of the pump outlets, a smaller pump suffices - and energy usage is reduced. The above table has been compiled by RV Books from makers' published data.

Chapter 14

Solar

There are various small-medium scale solar technologies, but all convert sunlight into electricity with efficiencies from 14%-21%. All are rated by a method best described as 'innovative', using Standard Operating Conditions that are far from 'Typical Operating Conditions'. The result is solar output less than that seemingly claimed.

What solar modules produce

Most solar cells dislike heat. Most vendors claim their heat loss begins at 25° C, but this misleads because that temperature relates to glass-covered black cells under a hot sun. According to the industry's data, at an ambient temperature of 25° C, those cells are at 47° C-49° C. They lose 5% of their output per 10° C increase in temperature and begin to do so at about 5° C. Thus, at 25° C, power output is down 10%, at 35° C it is down by 15%.

A further loss is caused by solar modules developing their maximum output at a higher voltage than conventional batteries can accept. It's like driving uphill in too high a gear. Unless you change down, or an automatic transmission does it for you, the needed output (at higher engine speed) is not accessible. Maximum Power Point Tracking (MPPT), included in the more sophisticated solar regulators, recovers part of this loss. (See Chapter 16).

Nominal Operating Cell Temperature

Unless your solar array is atop a mountain in a cold but sunny desert, your solar modules generate less than appears to be claimed. Module makers do reveal this but in terms called Nominal Operating Cell Temperature (NOCT) that mostly only engineers and technicians understand.

Translated, NOCT assumes an ambient temperature of 25° C, the sun at 48° C above the horizon and a breeze of 1 metre/second (3.6

km/h). Using the NOCT rating, a '120-watt' module in much of the world puts out about 87-watts, about 72.5% of that claimed: a bit more in cold parts and a bit less in hot parts.

Grid-connect systems usually incorporate MPPT regulation; hence that 85% of system output is about right. A system marketed as being '1.5 kW' rarely exceeds 1.2-1.25 kW.

How much sun?

Quantifying solar irradiation is covered in Chapter 1. Solar irradiation maps for Australia are in Chapter 25. These maps relate to mid-winter and mid-summer. In most places, the transition is more or less linear, but in some (like parts of New Zealand's north island) the change can be faster due to the rapid onset of cloud cover. Maps for other parts of the world are available from meteorological offices in the areas concerned.

As a reasonably valid generalisation, the highest annual solar input is between latitudes of plus/minus 15°-30°. By and large, areas closer to the equator receive less, mainly due to higher cloud/haze cover.

For stand-alone solar systems, one needs to minimise solar draw. However, as refrigerators draw less in winter, you need less power during these months (unless you have an unventilated fridge in a heated kitchen).

Tracking the sun

Having modules track the sun enables them to capture more sunlight early and late in the day. Tracking is most effective in winter at high latitudes, but progressively less otherwise. While common when module capacity was costly (or space at a premium), reliable tracking mechanisms were far from cheap.

Solar module capacity is now less than 10% of that then price. It makes sense (in most are areas) to accept some loss and increase solar capacity to compensate.

Ideally, have the modules facing true north and tilted at the angle shown in the section on solar module installation – Chapter 28 onward.

If intending to use one or more of the ultra-efficient reverse-cycle air conditioners for winter heating, consider increasing the tilt angle to as much as 50°-60°. This change makes a significant difference in winter and only a minor loss in summer. Do not be overly concerned that the modules face true north (in the southern hemisphere) or true south (in the northern hemisphere).

Errors of 5 to 7.5 degrees result in more input in the morning and less in the afternoon - or vice versa - but there is usually only a few per cent difference in daily input. Most energy is captured between 10 am and 2 pm.

Shadowing tolerance and loss

The term 'shadow tolerance', when used in connection with solar modules, is often misused. None work at all in full shade, they must have at least diffused sunlight. What is being referred to is the amount of output lost if *part* of the module surface is fully shaded, e.g. by a tree branch.

Shading most solar modules by more than 20% or so effectively shuts them down. A few solar modules have so-called by-pass diodes that partially assists. Even with these, however, a tiny shadow is likely to cause a one-third loss.

Figure 1.14. Significant loss of solar input by shadowing from nearby trees (that may have grown since the solar array was installed). Pic: solarpoweraustralia.com.au

61

Weather effects

It is rare to have no solar input, but total cloud cover may cut it by 80-90%. Heavy rain may reduce it to 5% or so, and smoke and haze from bushfires to zero input.

The highest input is generally obtained on sunny days with scattered light cloud. Then, if close to water (or a reflective surface), the light may be reflected upward, and down again. It is not unusual to have 20% or more increase for a brief period.

Buying solar modules

Solar equipment pricing is usually an integral part of a solar supplier's quotation. While rarely itemised, it is advisable to know who is the solar module maker.

There are only ten major solar module makers worldwide. These are (in alphabetic order): Canadian Solar, First Solar, JA Solar, JinkoSolar, Hanwha Q Cells, Longi Solar, Risen Energy and Trina Solar. It pays to ensure those quoted for are from any of the above.

If self-implementing your system, however, it pays to shop around. Many suppliers offer substantial discounts to those buying modules and batteries in bulk.

There is a minor trap when buying modules. Some are explicitly intended for grid-connect systems and are rated at voltages that are not multiples of 12 (volts). These are fine for stand-alone systems if used with a suitably compliant MPPT regulator (that typically accepts a wide range of input voltages), but rarely otherwise. Many buyers seeking bargain modules on eBay are caught out by this.

Chapter 15

Solar modules – voltage & current

Single solar modules vary in output from a few watts to 350-watts or more. Many solar-powered systems, however, need more output than a single solar module can produce. That increase can be achieved by using multiple modules.

The same number of solar modules can be interconnected in various ways. Depending on that required, you can increase amps (with volts remaining as is) or increase volts (with amps remaining as is).

Connecting two modules in series (i.e. 'end to end') doubles their overall voltage, but the current remains the same. Connecting two in parallel (i.e. 'side to side') doubles available current, but the voltage remains the same. Either way, the total available energy remains the same.

For systems up to 1000-1500-watts, it is common to use 12-volt solar modules in parallel to increase current. Connecting in parallel is done by interconnecting the solar modules' positive terminals, likewise the negative terminals. The resultant current and wattage is the sum of that of the individual modules.

Solar modules of different *wattage* can be paralleled as long as they are all of the same voltage. The output is the sum of their wattages. It is thus feasible to parallel connect a 12-volt, 100-watt module, and a 12-volt, 50-watt module. The combined output is 12-volts at 150-watts.

Increasing voltage

Once beyond 1000-1500-watts, the current (of about 85-125 amps at 12-volts) requires a bulky cable. It is better to have the solar array, and associated battery bank, running at 24-volts or more. You can do this by using 24-volt modules, or by connecting 12-volt modules in series, i.e. end-to-end. The resultant voltage is the sum of the individual modules' voltages. The current remains the same as if it were one module.

Series-connected solar modules must all be of the same, or very similar, current output. The overall current output is limited to that module of the least current output. Voltages may, however, be different. The total voltage is the sum of *each* module's voltage. To obtain 24-volts, it is thus feasible to series-connect 12 two-volt batteries, four 6-volt batteries or two 12-volt batteries.

Earlier owner-built stand-alone systems had the solar array running 72 or more volts. Non-electricians are now restricted to working on (or installing) anything that is over 60-volts DC. That voltage, however, is ample for a 24 or 48–volt battery bank.

Increasing voltage and 230-volts AC

To increase both current *and* voltage, have strings of series-connected solar modules in parallel. Where higher power is needed, have sets of the above with compatible paralleled inverters providing whatever is needed. It is feasible to have three-phase output - but that's licensed electricians' territory.

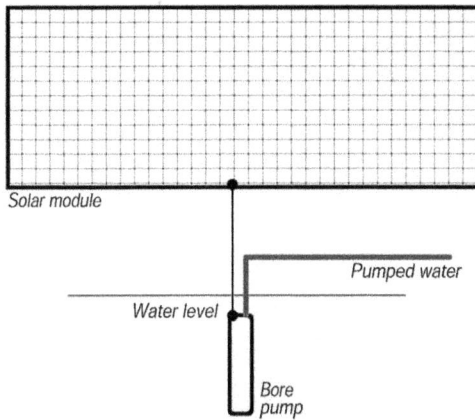

Figure 1.15. A basic system – a single solar module and pump. Pic: rvbooks.com

For basic stand-alone systems of 100-500-watts used perhaps for water pumping (Figure 1.15), it is simplest to use one or more 12-volt modules. A battery is only needed if pumping is required during periods of no sun.

Power Maximisers

Power is a measure of the rate at which work is done. It is measured in watts (one watt is the product of one volt and one amp).

Power as such cannot be self-increased, but in electrical form, at least, it can be manipulated to suit various needs. This ability is particularly useful for water pumping as many basic pumps require 15-20-volts – yet need to run from a nominally 12–volt solar module.

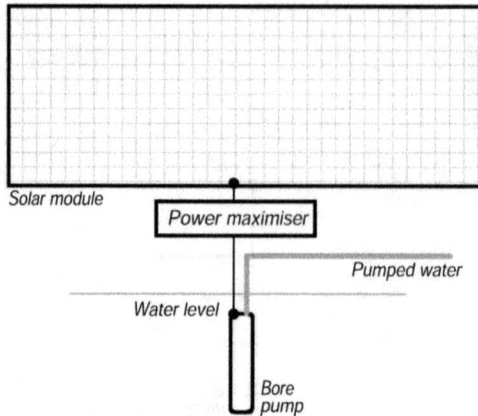

Figure 2.15. Here, a power maximiser improves pumping efficiency. Pic: rvbooks.com.au

So-called power maximisers work much as does a car's torque converter, it 'juggles' amps and volts to optimise that needed.

A typical system suitable for an outback cabin or a basic home is shown here (Figure 3.15) with four 12-volts, 100-watt solar modules, battery bank and charger plus an inverter to provide 230-volts AC. A standby generator provides back-up during extended periods of low solar input. In practice, it is likely to have far more solar modules.

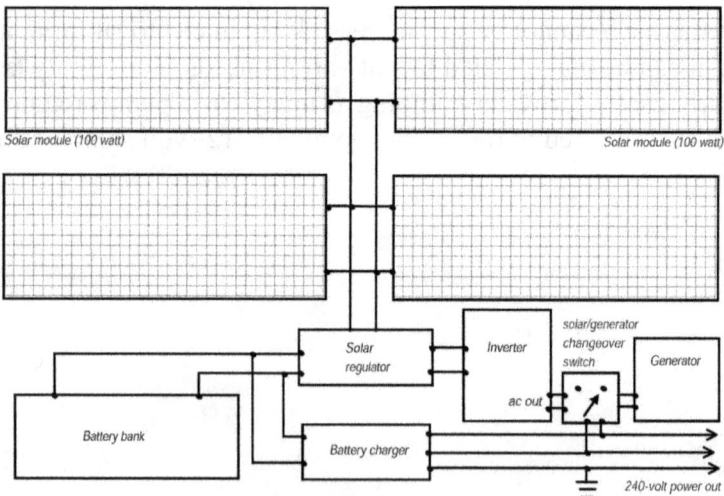

Figure 3.15. Basic cabin/small home system.

Figure 3.15 shows a typical system suitable for an outback cabin or a basic home. It has four 12-volts, 100-watt solar modules, battery bank and charger plus an inverter to provide 230-volts AC. A stand-by generator provides back-up during extended periods of low solar input. In practice, it is likely to have far more solar modules.

Solar output beware – it's not what it seems

A '100-watt' solar module should, by mathematical definition pro-duce 8.33 amps at 12-volts. It does not. If charging a battery at 12-volts, that module's actual output is about 5.9 amps (71-watts). At 12.8-volts it is 75.5-watts.

The solar industry defends this on the basis that it is a traditional in-dustry practice (but so was piracy in the traditional sea-faring trade). That rating, at a so-called Nominal Operating Cell Temperature (NOCT) rating, is rarely quoted on the packaging.

Solar reality is more or less 70% of that claimed promotionally. The actual output *is* revealed in technical specifications and often on a small panel on the rear of the module (Figure 4.15), but in terms that only technicians understand.

IRRADIANCE AND CELL TEMPERATURE	$1000Wm^{-2}$ AM 1.5 25 °C	$800Wm^{-2}$ AM 1.5 47 °C	MAX.SYS VOLT. 600 V
Pmax	120W	87W	SERIES FUSE
Vpmax	16.9V	15.2V	11 A
Ipmax	7.10A	5.74A	MASS
Voc	21.5V	—	11.9 kg
Isc	7.45A	—	

Figure 4.15. As can be seen in the third column, this '120-watt' module produces 87-watts at a cell temperature of 47° C – and an ambient temperature of 25° C. Pic:solarbooks.com.au

Chapter 16

Solar regulation

The voltage across a solar module varies with the intensity of the sunlight. That voltage must be controlled to safeguard connected appliances and to ensure associated batteries are optimally charged.

The crudest way of doing the former (but not the latter) is via 'self-regulating' solar modules. These have a limited number of cells connected that allows the module only to produce a maximum of 14.5-volts, thus making it challenging to overcharge the battery. In essence, they self-regulate in a manner akin to starvation being a 'self-limiting diet'. Their output is usually too low to more than half-charge a battery. If used without a solar regulator, however, they can occasionally produce voltage high enough to damage a battery.

The most basic solar regulator is a voltage sensing switch between the modules and the battery. That switch remains closed at low battery charge, opening every second or two as the battery approaches full charge. The cheapest solar regulators cost $35 to $100 or so and handle about 10 amps. They are reliable but charge inefficiently. They are only borderline acceptable for basic systems - but not recommended.

Sophisticated solar regulators cost $250 upwards. They control charging current as well as charging voltage. Most are programmable to handle input voltages of 12 to 36-volts at their rated current (some handle higher input voltages). All need programming for time, battery type, voltage and capacity. Programming is not complicated once the manual has been read a few times.

Conventional regulators have inherent losses, e.g. 12−volt batteries require 13.2-14.7-volts to ensure fully charging, but solar modules typically produce about 17-volts. The more basic solar regulators cannot access that voltage in between - resulting in a loss of that energy. This issue is addressed via the MPPT systems described below.

Maximum Power Point Tracking (MPPT) regulators

An MPPT regulator accepts input voltages from as low as 9-volts up to a typical 72 or so volts (some accept well over) and is programmable to charge 12, 24 and 48-volt batteries. It optimises the input by 'juggling' volts and amps thereby maximising energy input and the battery bank's needs at all times.

No longer costly, the MPPT function is particularly useful in cold places (where solar modules produce a higher voltage). It enables that higher voltage to be exploited. When voltage is low due to little sun, it increases the voltage at the expense of current, thereby enabling the battery to charge earlier and later each day. Another benefit is that it enables the output of the solar array to be at a high voltage. That high voltage enables lighter cabling: particularly worthwhile where the solar array is at some distance from the battery bank.

Figure 1.16. This Steca PR 1010 solar regulator handles 10 amps (at 12 or 24-volts). Pic: Steca.

Vendors typically claim 'gains' of 20% to 30%, but MPPT only recovers energy otherwise lost or not accessible (a likely 10% - 15%). No *increase* in energy captured is possible. The technology is usually built into up-market solar regulators integral with grid-connect inverters. It is also built into individual solar modules.

Up-market solar regulators include monitoring that shows what is happening to the battery and solar system. It is feasible to add this to regulators that have no readout, or where that existing is inadequate.

Figure 2.16. This Outback Power MPPT
regulator handles up to 70 amps.
Pic: Outback Power.

Chapter 17

Batteries

With stand-alone systems, a battery bank handles loads beyond the solar input's immediate ability to supply. The battery bank also stores excess solar energy for use outside daylight hours and during little sun. With grid-connect systems, there is already an escalating trend to include battery back-up.

The cost of solar capacity fell by 80% or so between 2010-2018: that of battery capacity rose by over 40%. The latter cost increase does not necessarily increase the overall system cost. Adding more solar capacity not only enables quick charging but also increases input during periods of little sun. It is thus feasible to maximise the now low-cost solar capacity and reduce high-cost battery capacity.

Battery types

Conventional batteries provide energy over time but, as the graph below shows, their lifespan is shortened if they are routinely deeply discharged. Their description (of 'deep cycle') can thus mislead.

If discharged overnight by 15% (about 85% remaining), good quality batteries withstand 4000-5000 cycles of such use. Rare deep discharges do little harm, but battery life is substantially reduced if done frequently. In essence, battery makers sell usable amp–hours. You can use a few for a long time, or a lot for a short time, or optimally in-between.

AGM batteries cost more than conventional lead-acid batteries, charge faster, and suffer less harm if deeply discharged many times. They are a good buy for remote cabins, and for people who would sooner pay more and forget that batteries are there. They require the associated solar regulator and battery charger to be programmed explicitly for such use.

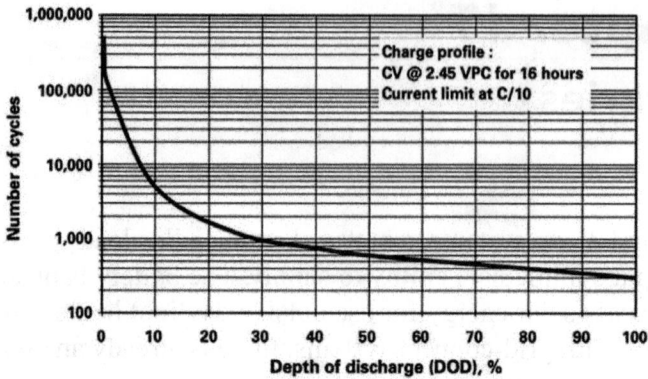

1,000,000

Number of cycles

100,000

10,000

1,000

100

Charge profile :
CV @ 2.45 VPC for 16 hours
Current limit at C/10

0   10   20   30   40   50   60   70   80   90   100

Depth of discharge (DOD), %

Figure 1.17. Deep cycle batteries - typical life span vs depth of discharge. Graph: Woodbank Communications Ltd mpoweruk.com

Lithium-based (LiFePO4) batteries are claimed to be ultra-rugged, able to be almost fully discharged without harm and to have excellent longevity. Their management and charging are more critical than for most other battery types, so it is advisable to buy both battery/s and charger from the same supplier.

In 2018 lithium batteries were about one third the weight, and size of other types of a battery of similar usable capacity, but three to four times the price.

Battery capacity

Do not assume big battery banks are better than small ones: it is like having several bank accounts for the same income. If energy is not there to save, increasing the battery capacity (alone) *increases* overall losses. Have only the battery capacity that can be fully charged on almost all days.

The amount of energy that can be drawn from conventional lead-acid and AGM batteries is related to the *rate* at which it is drawn. For example, where a 150 amp–hour capacity such battery has to supply the typical 150 amp draw of a microwave oven via an inverter for (say) 10 minutes, the battery retains its remaining capacity. It cannot, however, sustain that draw without its voltage dropping too low for the microwave oven to continue working. The full remaining capacity is still there, but not available under high rates of discharge. This limitation is not an issue with lithium batteries – even those of 10-20 amp–hour can release huge currents.

Interconnecting batteries

As with solar modules, batteries can be interconnected in various ways to increase voltage or current.

Doing this is usually necessary (at 12-volts) as single lead-acid, and AGM batteries larger than about 100 amp-hours are too heavy to move safely. The method of interconnection is similar to that of solar modules. Still, there are additional restrictions, including that it is not advisable to interconnect new and old batteries.

Current is increased by parallel connection, but care is needed to ensure all batteries receive an identical charge. Some people maintain that parallel connection should be avoided. However, the Exide Corporation (one of the world's largest battery makers) advises that as long some precautions are taken, it is fine to parallel connect up to ten batteries. Exide adds that, if paralleled, all batteries must be of the same voltage, but can be of various amp-hour capacities. The total current output is the sum of the individual batteries' current outputs.

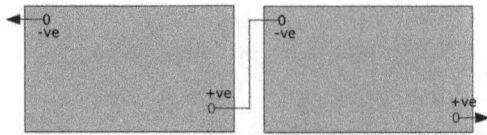

Figure 2.17. Two series-connected batteries. Their voltage is the sum of each battery. Were these 6-volt (100 amp-hour) batteries their combined output would be 12-volts (100 amp-hour).

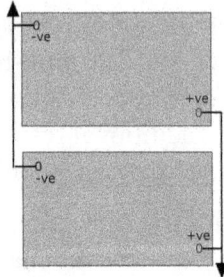

Figure 3.17. Two parallel-connected batteries. Their voltage is the sum of each battery. Were these 6-volt (100 amp-hours) batteries their combined output would still be 6-volts, but the capacity a now total of 200 amp-hours.

Voltage is increased by series connection: positive to negative, positive to negative. Four 100 amp-hour 12-volt batteries connected in series produce 48-volts at 100 amp-hour (4800-watt-hours).

Any number of batteries can be connected in series as long as they are all of the same amp-hour capacity. With a series connection, voltage is additive. Current remains as for a single battery.

Large systems that need (say) 48-volts at high currents may use parallel-connected strings of series-connected batteries. In Figure 4.17 below, (a) shows four series-connected 12-volt, 100 amp-hour batteries, (b) shows the same batteries connected in parallel and (c) shows two paralleled strings of series-connected batteries.

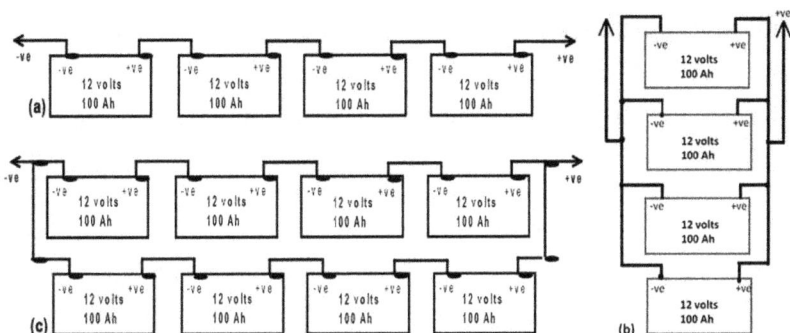

*Figure 4.17. Battery connections: (a): series connection – 48 V at 100 Ah
(= 4800-watt-hours) (b): parallel connection – 12 V at 400 Ah (=4 800-watt-
hours), (c): series-parallel connection – 48 V at 200 Ah (9600-watt-hours).
Pic:solarbooks.com.au*

As with solar modules, there is no 'magic bullet' way of interconnecting batteries to obtain 'more power.' The same number of similar batteries always stores and delivers the same watt-hours.

Battery location

In 2018 Standards Australia, a press release relating to the an-
nouncement of the draft *voluntary* standard AS 5139. This applies
to batteries installed in a fixed location whose voltage is at least 12-
volts and whose energy storage capacity is at least 1 kilowatt-hour.

For example, a wall that is required to meet an FRL of 120/60/30
means that the wall must maintain structural adequacy for 120 minutes, integrity for 60 minutes and insulation for 30 minutes, as tested to AS1530.4. noted that 'in considering the fire hazards associated with some of these systems, the draft contains provisions that
exclude certain battery systems from being installed inside domestic homes. They may be installed externally and adjoining to domestic homes provided certain fire related safety measures are met.

While 'certain battery systems' were not defined it almost certain
they include lithium-ion. The fire related safety measures include a
'detached building or structure where the battery system shall be
housed in a purpose-built room or enclosure rated at the FRL of a
minimum of 60/60/60.'

The term 'FRL of a minimum of 60/60/60' refers to the ability of a
structure to not fall apart during a fire, to restrict the passage of

flame and hot gasses, and to limit the heat on the non-exposed (or non-fire side) of a fire resistant separating barrier 60 minutes. .

For example, a wall that is required to meet an FRL of 120/60/30 means that the wall must maintain structural adequacy for 120 minutes, integrity for 60 minutes and insulation for 30 minutes, as tested to AS1530.4.

In essence it advises to locate lithium and similar high energy density batteries in a dedicated enclosure that meets FLR 60/60/60 requirements.

Chapter 18

Battery charging

A battery is charged by applying a voltage across it that is higher than it has at the time. The greater that voltage difference, the higher the charging current, and the quicker the battery charges.

Subject to the above, lead-acid deep cycle batteries can usually be charged at 15% to 20% (in amps) of their amp-hour capacities. A 100 amp-hour such battery can be safely charged up to 20 amps or so. Ideally, the charger should be capable of charging at not less than 10% of that bank's amp-hour capacity. A 500 amp-hour battery needs a 50 to 75 amp charger.

AGM batteries can be charged at higher rates but charging at much over 15% is rarely required for home/property systems. The same applies to the now less commonly-used gel cell batteries.

Lithium batteries withstand very high charge and discharge rates. They require specialised battery management systems to ensure equal cell charging. These systems are essential - but not always supplied as part of the battery pack. Their charging should be discussed with the battery vendor.

Do not use cheap battery chargers - they tend to overcharge and wreck batteries. This effect is readily identifiable by battery vendors and leaves you without redress.

Good chargers are not cheap: expect to pay about 20% or so of the price of the batteries they charge. The charger should be programmable for conventional lead-acid, gel cell and AGM batteries. Buy only from specialised electrical and solar suppliers.

Most pre-1990 battery chargers used transformers to reduce the 230-volt AC input to that required. These chargers were reliable, but only about 70% efficient. Today's switch-mode transformer-less chargers cost much the same but are smaller and lighter. Some are over 95% efficient. Switch-mode chargers (Chapter 18) may not run satisfactorily, or at all, from some generators.

The battery charging area

A secure and cross-ventilated area must be provided for the battery bank. Details of the vent sizes required are in Chapter 18.

Figure 1.18. The author's previously-owned battery rack was self-built from industrial galvanised Dexion has bracing gussets at every corner and shelf.
Pic: solarbooks.com.au.

Large batteries are heavy and cumbersome. The 12-volts, 230 amp-hour units (Figure 1.18) totalled over 1.5 tonnes. The racks not only needed to support the weight but had to be braced to ensure they did not collapse sideways or diagonally.

Figure 2.18: A warning sign like this should be located on or close to the battery charging area's door.

The battery charger should preferably be located in a separate but adjacent secure area. Details of battery and battery charger installation are in Chapter 28.

Chapter 19

Energy monitoring

For all but the smallest systems, you need to know how every part is working, particularly the battery bank's ongoing state of charge. Monitoring checking voltage alone is not good enough because deep cycle batteries react very slowly to change, especially while recovering from heavy loads. An accurate voltage measurement requires the battery bank to have rested off charge, and with no load, for about 72 hours.

Several companies market top-quality energy monitors that typically also check and show current draw, ampere-hours consumed, the remaining time available at the current rate of discharge. Many are programmable to turn off non-critical loads or to run a generator when needed.

They all work much as you keep account of money. Note what comes in, deduct what goes out – and deduct the bank's ransom for using your stored money. The result is what you have left.

Figure 1.19. Victron energy monitor.
Pic: Victron.

All but the cheapest solar regulators have some monitoring inbuilt. Where, there is none, or where what you is too complicated, you can add an energy monitor, such as that shown in Figure 1.19. (See also Chapter 24).

Monitors show the charge coming in, the total for the previous day, and the energy being used. Most have facilities for turning things on and off under pre-settable conditions, e.g. 'start the generator if battery voltage drops below whatever, and stop it again when the batteries are 98% charged'.

Where energy monitoring is included within the solar regulator, the current drawable from the system is often limited by that regulator's rating. This limitation can be overcome by adding a current shunt.

Figure 2.19. Typical current shunt.
Pic: rvbooks.com.

A current shunt monitors all current into and of the battery. That information may be displayed on the solar regulator's screen; or a separate readout. Most displays also store monthly data that is handy for subsequent fault finding.

Some solar vendors install monitoring accessible only from their premises so they *alone* can fix faults. *Insist* on data being accessible on-site as many faults are trivial and readily owner-fixed.

Low solar input, for example, often requires only that solar modules be cleaned, a shadow-causing tree branch trimmed or a battery lead re-tightened.

On-site monitoring may indicate what to check. Monitoring units need programming, but while the instructions may at first appear daunting, usage is usually easy once read a few times.

Chapter 20

Generators

While possibly optional with systems that have LP gas fridges, a generator will sooner or later be required for a home or property system. Its size and type depend on how you wish to use it. A generator can be an integral part of the system to provide power if something goes wrong, or to cope during spells of reduced sunlight, or to run occasional extra-heavy loads.

For occasional use for cabins, RVs and small homes, the petrol-powered units, such as the quiet Honda/Yamaha/Suburu inverter generators, are cheaper, smaller and quieter.

Figure 1.20. This Honda 3000is petrol-fuelled inverter generator can produce a continuous 2400-watts.

Be wary of ultra-cheap petrol-engined units.

Buying a cheap generator is a severe risk for buyers who believe that Lexus performance can be bought at Lada prices. Apart from lacking voltage regulation, when one of these run out of fuel, it sputters to a halt, generating voltage spikes that can damage anything connected to it.

The high-quality units are at their most economical when running at about 60% of their rated output. Their continuous use is typically 80% of their rated output.

Diesel generators for long term use

As noted above, petrol-powered inverter generators are intended for occasional use. They do that well and reliably but have a limited lifespan. They are not intended, nor suitable, for long-term none-stop use.

For frequent long-term use, it usually pays to buy a top-quality diesel-powered unit. Such generators cost three or more times that of a petrol-driven unit of similar output. They will, however, use 35% or so less fuel and are likely to last several decades.

There are two basic types. That for 50 Hz (cycles per second) output run at either 1500 rpm or 3000 rpm. The 1500 rpm units cost more but may outlast their owners. The 3000 rpm units cost less. Their reliability/longevity is acceptable, but they do not rival their slower speed brethren.

Diesel-powered generators have the further advantage that their high compression ratios necessitate heavy flywheels that have a great deal of momentum. As long as the generator is running, the flywheel's momentum supplies energy to assist in starting heavy loads such as a large air compressor. A good diesel generator may thus start a load that is way beyond the capability of a similar-sized petrol-fuelled unit.

For frequent long-term use, it usually pays to buy a top-quality diesel-powered unit. Such generators cost three or more times that of a petrol-driven unit of similar output. They will, however, use 35% or so less fuel and likely to last several decades.

A diesel unit is less suited to occasional use as the fuel tends to fungal build-up, but that is substantially reduced by additives (obtainable from truck fuel stations).

Battery charging from a generator

Some generators have a 12 or 24-volt DC output. Even if labelled 'battery charging', that output is a typical 13.65-volts at a maximum of 8 amps. Except to partially charge a car starter battery such voltage is too low. Instead, use a high-quality battery charger driven from the generator's 230-volt output.

A few specialised generators (both petrol and diesel) generate 12 or 24-volts DC only. They eliminate the losses involved within conventional transformer type chargers (that need a generator with output about twice that of the charger's output). Dc generators still have a place in specialised applications. A generally better approach, however, is to use a conventional 230-volt AC generator and a switch-mode charger.

Auto-start/stop

Most generators can be set up to start automatically when the battery voltage is below a preset level and then stop at a higher preset level. Facilities for doing this are usually built into the solar regulator or energy monitor, but as noted in the section on installation, setting this up may need specialised help.

Generator noise

Unless silenced, diesel-power generators (in particular) are noisy beasts. Even small ones can be heard from a kilometre or more away. It is possible to quieten them to at least acceptable levels. Details are included in the section on installation (Chapter 28). Some manufacturers make enclosed silenced units.

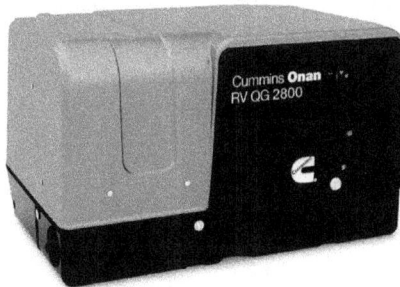

Figure 2.20. Cummins Onan RV QG 2800 weighs 57 kg (125 lbs). It meets the USA's National Park Service sound level requirements of 60 dB(A) @ 15 metres (about 50 ft).

Generator size

For stand-alone outback systems, there is a good case for a unit large enough to provide backup in the event of a major failure (such as a direct lightning strike on the solar modules). Also, in the event of a long period of dense cloud, a fair-sized generator can supply power while also recharging the battery bank.

It is better to have a generator that is too large than too small - but do not go overboard, or you'll spend money unnecessarily on fuel. Generators are most economical when running on 50%–80% full load.

Generators and switch-mode chargers

Most switch-mode chargers are protected against spikes and other electrical nasties that can prevent them from working correctly (or at all) from cheap generators.

In cases where the charger does not run from your generator, re-dress is close to impossible to obtain. Each vendor blames the other, not least because the generator probably *can* run other loads, and the charger is likely to work well from the grid supply. The problem usually *is* the generator but proving it is close to impossible.

Avoid this situation up-front by buying both units from the same vendor and stressing, in writing, that each unit must work with the other. This issue is mostly encountered with the massively polluting two-stroke $99 petrol generators that grey nomads sadly tend to buy. It is rarely encountered with diesel generators.

For general property use, there is a good case for an additional por-table 5 kVA or so petrol unit. A diesel unit is less suited to occa-sional use as the fuel tends to fungal build-up – but that is substan-tially reduced by additives (obtainable from truck fuel stations).

Chapter 21
Alternative power - wind & hydro

In the late 1800s, Sir Charles Wentworth Dill (not Mark Twain) commented that there are three kinds of lies: 'lies, damned lies, and statistics'. While not implying a parallel, it may pay to bear this in mind when evaluating some wind industry promotion.

While they work well on boats, wind power systems with propellers smaller than two to three metres are virtually useless in urban areas.

The main drawback is that their output is proportional to the square of the propeller's diameter and the cube of the wind speed. If wind speed halves, output drops to one eighth. If wind speed doubles, it increases eight times.

Promotional material typically shows the current output (in amps) from wind speeds of 10 km/h. It rarely, however, discloses that the current is at too low a voltage to be of value, at least for 12–volt battery charging.

Few small wind turbines have usable output until wind speed exceeds a (rare) constant 20-25 km/h. Many produce their claimed rated output only just short of the wind speed that blows them apart.

One supplier states: 'the average wind speed needs to be above 18 km/h 'to make installing a wind turbine worthwhile'. The average yearly constant wind speed, even in the windiest parts of Australia (including Cape York, Albany and Geraldton) is less than 15 km/h. Across most of Australia, it is far less.

Local councils tend to be paranoid about the required high masts, and neighbours rightly object to wind generator noise. The units are hard and potentially dangerous to erect yet require ongoing maintenance more safely done at ground level. They may require mechanical restraint or even taken down in heavy winds.

On top of that, the associated industry is unregulated. In most countries, related performance standards are non-existent.

Larger wind generators

Due to the square law involved, wind power begins to be valid with propellers of about five metres diameter and above. These are well worth considering on isolated windy properties but need to about 35 metres above the ground. At times they may develop more energy than can be used or stored in batteries. That energy, however, but could also be used for pumping water. The bigger units need skill and experience to take up and down.

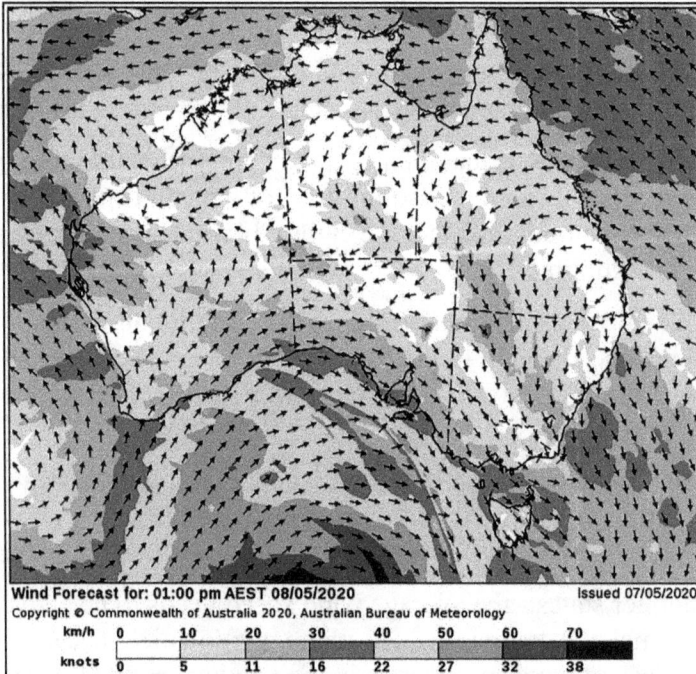

Figure 1.21. Windspeed map for Australia.
Pic: Australian Bureau of Meteorology.

Micro hydro-electric

Where there is a reliable supply of water and sufficient pressure (as in parts of New Zealand), small hydro-electric turbines are cost-effective power generators.

The systems pipe water from as high as possible and direct it to impinge on a water wheel (turbine) at high pressure. The turbine is usually coupled to an electricity generator. Output depends on water pressure and volume. For small systems, the former is a priority.

The site requires a substantial fall over a short distance. The water wheel needs locating next to the generator, as it is cheaper to run electrical cable than the large diameter water pipes needed to limit energy losses over a more prolonged fall.

Figure 2.21. This micro-hydro system built by Anastase Tabaro provides power to households in his village.
Pic: European Press Agency.

Turbine types

Different types of water turbines can be used depending on the head of water, flow rate and a few other variables. One, (the Crossflow turbine), works well at heads ranging from one to 100 metres.

Figure 3.21. The Crossflow turbine. Pic: pumpfundamentals.com

With a water head of 50 metres or more, a traditional Pelton wheel can be used. Where there is ample flow but little pressure, propeller-type turbines work well. Minimal systems may use centrifugal pumps run back to front as generators.

A small hydro-electric plant is not expensive if major earth-moving is not required. Experts in this field advise that the cost is typically one third that of solar energy.

Ram pumps

Ram pumps (described by alchemist Robert Fludd in 1618) are believed to have been used since the 1500s. They need no external source of energy than flowing water.

Figure 4.21. A ram pump in Africa. Pic: RNZ.

These pumps are truly basic but ultra-reliable. Their downside is that only 10% or so of the water flow through the pump is forced up the delivery pipe. The rest is lost via a release valve.

Ram pumps are ideal for pumping water from a free-flowing stream but are extremely noisy. They need locating kilometres away from neighbours.

For each metre of the initial fall (say 30 metres), a ram pump lifts it about seven times (to 210 metres). Alternatively, the water could be pumped horizontally.

Geothermal space heating

Geothermal energy works much as does a reverse-cycle air conditioner except that, instead of drawing in outside air, it operates by circulating a liquid through buried piping. Such piping extends laterally if space allows, or if not, up to 100 metres or so downwards.

Figure 5.21. Typical geothermal (vertical) installation.
Pic: thehydronicsteam.com

The earth's temperature close to the surface varies from as low as 7° C to 21° C, (typically 14° C in much of coastal Australia). This constant temperature can be used for cooling in summer. In colder areas, it can be used to assist heating in winter.

Currently available units produce 3 to 4 kW of heat for every 1 kW of energy drawn. It is early days for this technology, and its efficiency is bound to improve yet further. Its main benefit for heating is where ambient temperatures are low, and vice versa for cooling.

The technology was pioneered in Iceland over 100 years ago when Stefan Jonsson used it to heat his farm. Other farmers then did likewise. The first public building so heated (in 1930) was a school in Reykjavik. This system worked so well that geothermal heating was then used for the National Hospital and 60 private homes. Over 90% of Iceland's homes are heated this way.

Geothermal technology is being used in North America and Northern Europe, including in underground train stations in Switzerland and the UK. According to Australian government reports these systems as capable of producing less environmental harm as any other alternative space conditioning technology currently available.

Passive and direct solar heating

Developed initially in Denmark, passive and direct solar space heating design takes advantage of and optimises a building's site, climate, and materials to minimise energy use. In its most basic form, it is akin to blowing in hot air from an adjacent greenhouse. Variants use solar hot water modules. Some hybrids heat both air and water.

Early vendor claims for this technology often specify only the temperature increase, but this has no meaning unless the volume of air heated is known. The heat output is unlikely to vary much from brand to brand.

Energy efficiency first

Before adding solar to your new home design or existing house, remember that energy efficiency is the most cost-effective strategy for reducing heating and cooling bills. Choose building professionals experienced in energy-efficient house design and construction and work with them to optimise your home's energy efficiency. If you're remodelling an existing home, ensure cost-effective energy efficiency improvements are a priority.

Site selection

A passive solar home is heated by the sun shining through north or south-facing windows. The sun heats-up materials that can store solar heat as so-called thermal mass. That thermal mass also absorbs heat from warm air in the house during the winter.

Thermal mass (typically brick, concrete, stone and tile) cools by absorbing the sun's heat throughout the summer. Functional thermal mass materials are more efficient at storing heat. Masonry, however, has the advantage of being both a structural and external finish.

In well-insulated homes in moderate climates, the thermal mass inherent in the external and internal walls may eliminate any for additional thermal storage.

If you're planning a new passive solar home, a portion of the sunny side must have unobstructed sunlight. Take into account likely changes to the sun falling on wherever you plan to locate solar modules. Bear in mind that small trees become tall trees and that a new adjacent structure can block your home's access to the sun.

In some areas, zoning or other land-use regulations protect landowners' solar access. If solar access isn't protected in your region, look for a land that is deep from north to south and place the house on the far end from the sun.

Chapter 22

Alternative energy storage - fuel cells

Conceptually part generator/part battery, fuel cells do not burn fuel to produce heat: they generate electricity via an electrochemical reaction between a fuel and an oxidant. Excepting for a by-product emitted as a small amount of pure water, fuel cells are virtually pollution-free and also silent.

Some fuel cells use hydrogen as fuel and oxygen as the oxidant, but more commonly and less efficiently, by producing hydrogen within the unit from LP gas, natural gas, methanol, diesel, etc.

While some heat is generated in the electrochemical reaction/s, the cells use only a fraction of the fuel of a conventional generator doing the same work. According to the US Department of Energy, fuel cells are 40%–60% energy efficient - or more if the heat internally generated is also utilised.

Fuel cell history

The fuel cell concept originated in 1843. There was little meaningful development, however, until 1955. Then, General Electric's Willard Thomas and Leonard Niedrach made some progress. The results were further developed by GE and NASA and during the Gemini space project. Pratt and Whitney improved the design to ensure longer life and supplied fuel cells for the Apollo and Skylab projects.

Each Apollo craft had three cylindrical units - each 570 x 1120 mm and 109 kg. Each produced 560-1400-watts. The later Space Shuttle cells were 380 x 1015 mm, and 116 kg and produced 7-12 kW.

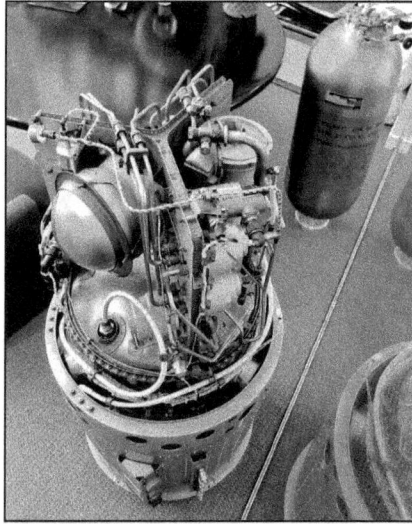

Figure 1.22. Apollo fuel cells.
Pic: Scott Schneeweis collection.

Commercial development

Large scale commercial fuel cells of a typical 100 kW upward have been available since the 1980s. Work has been underway to produce small scale (0.5 kW-5.0 kW) units for many years. Some are commercially available today.

There are many different types of fuel cells. A few run on pure hydrogen and run at relatively low temperatures. Most, however, use hydrocarbon fuels, such as natural gas.

A Korean research team has recently (2020) developed a ceramic fuel cell that can operate from butane. It is expected to be usable in portable and mobile applications such as electric cars, robots and drones. (Previously, ceramic fuel cells had only been considered for application to large-capacity power generation systems due to their high-temperature operation).

Fuel cell technology is developing fast, with any number of companies worldwide actively working on producing viable products for large scale production. The global market is enormous as it offers an almost pollution-free source of distributed electrical power - particularly needed in third-world countries. They are already being used in vast numbers to power hybrid and electric cars.

Fuel cells will *inevitably* become silent and clean replacements for conventional generators. They may eventually become a viable alternative to grid-connect generator back-up.

How fuel cells work

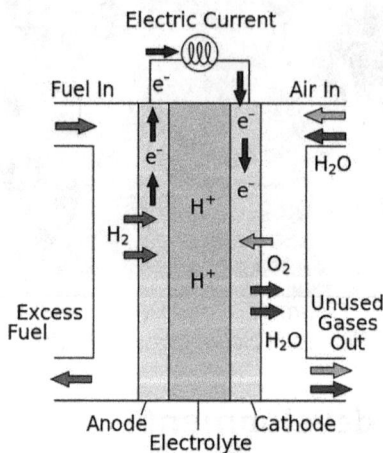

Figure 2.22. How a typical fuel cell works.
Pic: Qoncious.com

A fuel cell module's essential operation is battery-like. Each cell produces about 0.7-volts. Cells are connected in series to produce whatever voltage required.

Hydrogen is fed to the anode plate and oxygen (from the air) to the cathode plate. The hydrogen is split into positively charged protons, and negatively charged electrons, via a platinum catalyst.

Protons flow to the cathode via an internal polymer electrolyte membrane. Electrons flow to the cathode via an external circuit - thus providing usable electrical energy. The re-united protons and electrons combine with oxygen at the cathode.

Chapter 23

Inverters

Inverters convert direct current from a battery (or, with grid-connect, directly from solar modules and inverter) into mains-like alternating current. There are three types of inverter: square-wave (dating back to the 1970s those not already taken to the tip should be), modified square-wave, and sine-wave.

The cheaper, modified square-wave inverters may run older appliances but not necessarily appliances made since 2010 or so. Some appliances, particularly laser printers, are likely to be damaged by them.

Sine-wave inverters produce cleaner electricity than the grid supply. They can be relied upon to run any appliance within their rating. Only *these* should be considered for general home and property use.

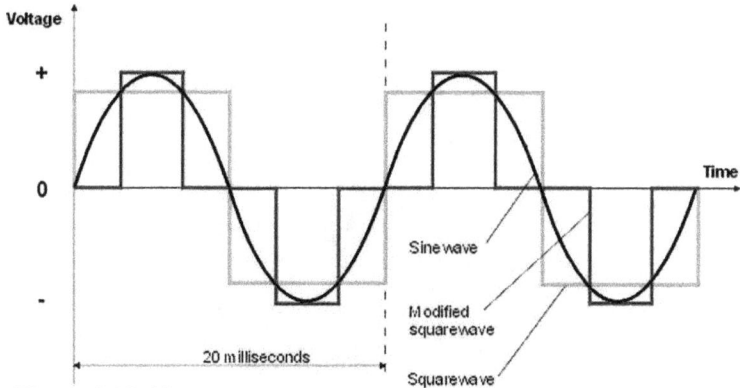

Figure 1.23. How voltage and current change (with respect to time) for a pure sine wave, modified square wave and square wave inverter. Pic: source unknown.

Silicon camels

Iron-cored transformer-based inverters are like silicon camels. They carry normal loads forever, accept minor overloads for up to half an hour and extra heavy loads for 10 to 20 seconds. They rest before restarting if severely overloaded. Such overload capacity is useful as many electrical devices two to four times their running current while starting. The downside, however, is that efficiency rarely exceeded 90%.

Later switch-mode inverters are smaller, lighter, and cheaper per watt, but most have next to no overload capacity. Some are derated for loads exceeding a minute or two. These inverters must be selected such that they can cope with the starting current drawn by many appliances.

Stand-alone inverters

The simplest inverters are stand-alone units, but most of these are limited to supplying around 500-watts. They usually have power outlet sockets on their front panel. Appliances plug directly into those sockets. To install, all that is needed is to run the usually supplied leads to the associated battery.

These units must not be connected into fixed mains wiring.

Figure 2.23. Inverters, such as this Projecta unit that have power outlet sockets built into the actual unit, may legally be self-installed but must not be connected to any fixed wiring. Pic: Projecta.

Inverters are generally limited to about 1200-watts at 12-volts, to 2400-watts at 24-volts, and 4800-watts at 48-volts. Some can be paralleled, e.g. twinned 2400-watt units can supply up to 4800-

watts. Such 'twinning' also ensures that if one fails, the other's output is still available. Figure 3.23 shows an example.

Figure 3.23. Four paralleled Outback Power inverters provide 14.4 kW. Pic: Outback Power.

Inverters for fixed wiring

Inverters intended for permanent installation have no inbuilt outlet sockets. They must be installed by a licensed electrician following the mandatory conditions of AS/NZS 3000:2018.

These inverters are made in various sizes. Some modular units (Figure 3.23), can be paralleled to provide larger capacity or three-phase output. Not all can do this so, if that may be needed in the future, check the capability before buying.

Figure 4.23. This Victron 24-volt inverter/charger provides 230-volts at 3 kW, and 6 kW surge capacity, plus up to 70 amps for battery charging. Pic: Victron.

Grid-connect inverters

Grid-connect inverters are much the same as other inverters - but with one major exception. Much as a surfer synchronises with a wave, the inverter's alternating current output must synchronise with the electricity grid. That synchronisation is ensured within the inverter.

Battery backed grid-connect

In order to protect electricity staff working repairing electricity supply cables etc. presumed to be 'dead', grid-connect solar systems *must* include a 'break before make' relay. This relay isolates power to and from the grid if the grid supply fails, or is shut down.

With many earlier systems, this left the user without power. That can now be overcome by specifying a specialised inverter at the time of the original installation. It can also be done via a separate inverter and charger (or an inverter with an inbuilt charging function), and an associated battery bank.

Battery back-up is increasingly becoming an integral part of a grid connect system (and retrofitted to existing systems). Such back-up enables excess input to supply the home's needs during peak supply periods. This system also benefits grid-connect users who are away from home during the day.

Victron's Energy MultiPlus range of inverter/chargers combine a sine-wave inverter, battery charger, and high-speed power transfer. Several such units can be paralleled for increased inverter and charger output. Multiple units can be connected to provide three-phase power.

Chapter 24

Energy auditing

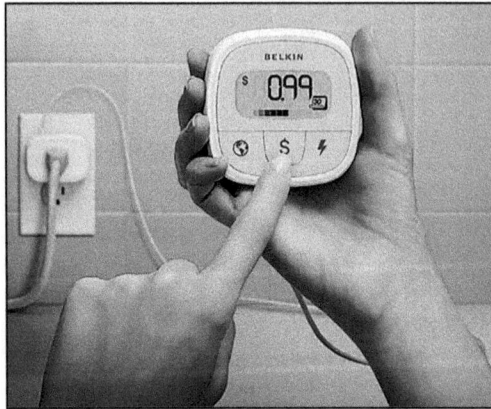

Figure 1.24. The Belkin Conserve Insight shows watts and also cost per kWh. Pic: Belkin.

An energy audit takes stock of all your electrical lights and appliances. It establishes how much energy each draws and for how long, on average, each is used per day. It highlights unexpected areas of waste and enables a comparison with known national averages and the generally lower usage of existing solar users. It is easy to do this yourself by using an energy meter, but be wary of cheap ones: many mislead when measuring low wattage loads.

Existing usage

Following the example in Table 1.24 list everything electrical you intend to use (in a copy of Table 2.24). For each item, note the amount of energy the item consumes. This information is usually embossed on that item or printed on an attached panel (Figure 1.24). It is also generally included in makers' specifications.

Consumption is usually shown in watts, but electric motors and microwave ovens may indicate wattage as a measure of work done, not the energy expended in doing so. Pump motors may show P_2, (i.e. the power the motor develops). Or they may show P_1 (i.e. the power the motor draws on full load). Ideally use that P_1 data. If wattage only is shown it is odds on that this is P_2 (power developed). If so adding 50% to that P_2 Figure, while not exact, is usually close enough.

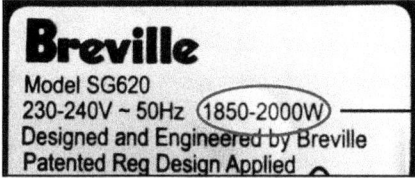

Figure 2.24. Most electrical devices have a label such as this that quotes energy usage (in watts).

If it is not feasible to establish the consumption, use the mid-range (where applicable) of whatever is shown in Table 1.24. This list is not exact, but errors more or less cancel out. Assess how long each device is likely to be used. Use probabilities, not exceptions.

Multiply each device's energy draw in watts by the probable hours used each day. The result is the energy draw in watt-hours per day. It is more convenient to show this as (say) 12.5 kWh/day, rather than 12,500-watt-hours per day.

The average energy draw of a three-bedroom grid-connected (but non-solar) urban home (in 2020) varies from 10-23 kWh/day. There is no official data for stand-alone solar installations. Anecdotal evidence, however, suggests it is less than that of grid-connected homes. If so, it is probably because their owners become more conscious of unneeded lights and TVs left on all day.

ITEM	Consumption (in watts)
Air-conditioners (2.5 kW)	600-1000 Most made before 2004 draw at least 750-watts.
Blankets	60-120
Blenders	400
Breadmakers (per loaf)	350-450
Clippers	35
Clothes dryers	1000-2500
Coffee grinders	75
Coffee percolators	500-600
Computers – desktop	400-500
– laptop	70-100
Computer printers – bubble jet	120
– laser	900-1000
Dishwashers	1000-3000
Fans (small)	30-40
– pedestal	75
– overhead	100-150
Food processors	450-550
Frying pan – electric	1200-2200
Fridges*	100-250
Hairdryers	500-1250
Infra-red grills	1500-2000
Irons	1200-2200
Juicers	350-450
Kettles/jugs	2200
Lights	
– compact fluorescent (typical)	5-22
– white LEDs	3-12 (extra-low voltage)
Microwave ovens (800-watts) – normal	1350
– convection	1800
Ovens	2500-5000
Radios	15-20
Televisions – 60 cm	18-50 (a few draw up to 75-watts)
– 80	40-60 (a few draw up to 100-watts)
– 100	40-110 (a few draw up to 150-watts)
– 120	60-130 (a few draw up to 170-watts)
– 140	75-175 (a few draw up to 250-watts)
Space heaters	1500-2000
Vacuum cleaners	750-1500
Washing machines	200-600
Waste disposal units	500-750
Water pumps – 12-24-volt DC	50-100
Water pumps – 230-volt AC	>750

*Table 1.24. An example of typical energy use. * Most older fridges continuously cycle on and off. The energy draw shown here is for the 'on' cycle. It is affected by the thermostat setting, ambient temperature, installation and usage pattern. The on/off ratios may vary from eight hours on-time a day – to continuous.*

Item	Watts	Hours /day	Watt-hours /day
TV	200	2.5	500
DVD	30	1	30
Total			

Table 2.24. Use this as a template to compile your version of Table 1.24

Chapter 25

Scaling stand-alone systems

This Chapter shows how to assess various sizes and capacities to ensure that the resultant system is adequately and proportionately scaled.

As outlined in the previous pages, note the amount of energy each appliance uses, and check against the specialised coverage in earlier sections of this book. Look particularly for any item that draws a great deal of power but is used only briefly and then rarely, such as an arc welder or big air compressor. These are better run from a generator as it makes no sense to grossly oversize any costly part of the system to cope with rare high usage. Study your energy audit carefully. Rework it if necessary and compare it against the example systems in Chapter 31.

Available irradiation

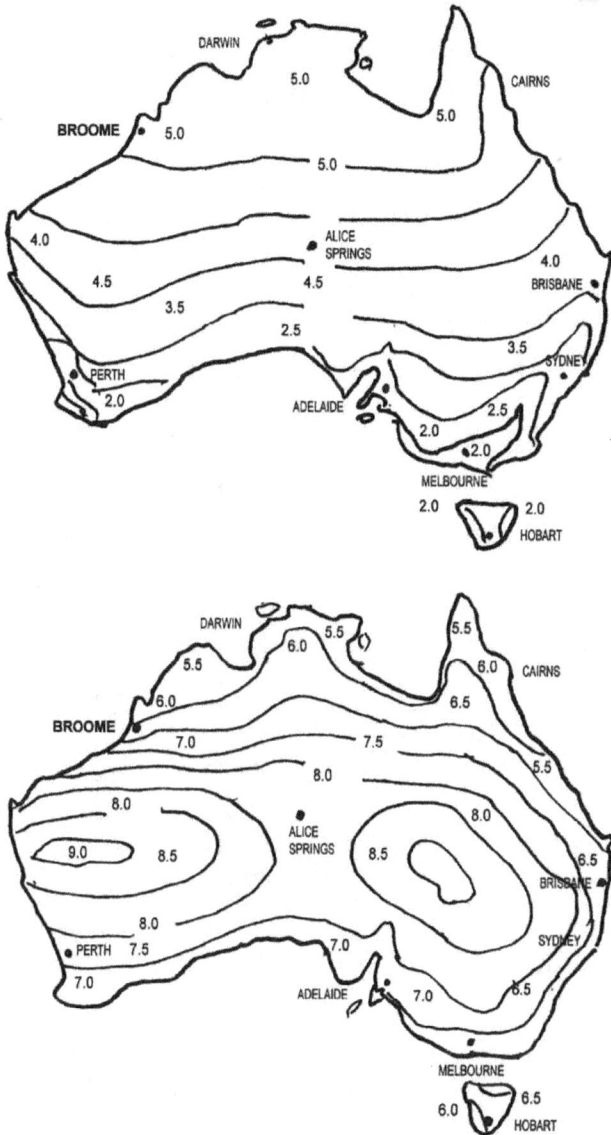

Figure 1.25. The above two maps are based on 10-year running averages.
The upper map relates to July, the lower map to January. To use, multiply
your nominal solar array wattage by the PSH for your area. Your available
solar wattage is about 70% of that amount. These maps are based on the
Australian Bureau of Meteorology data.

Using the solar above, check your summer and winter irradiation. Most places have a more or less smooth change between seasons, but this cannot be assumed.

Solar charts can be obtained from meteorology bureaus. Their average data is accurate, but there are usually substantial daily variations.

Note the relevant PSH (Peak Sun Hours) and multiply that by 70% of the solar modules' nominal capacity in watts. The result is the total watt-hours you are likely to average each day. For areas that in mid-summer have a typical 6.5 PSH, a nominally 100-watt solar module's probable daily input is thus 70-watts times 6.5 PSH = 455-watt-hours.

Needed solar capacity

A minimum yearly average of 2.5 to 3 PSH enables solar to be economically viable. Adding 25% more solar capacity than needed for everyday use (and cutting back on non-essential usage) gets you through periods of less than the usual sun. As a very rough guide, it is realistic economically to scale the system to provide 95%–97% from solar alone (except in the Melbourne area - where 85%–90% is realistic).

It's worth doing a few sums to work out the relative costs of adding more solar or buying increasingly grid-power. The balance is shifting in favour of more solar, not least because solar module cost fell 80% or so in price from 2010-2018 and still falling.

The decision may be influenced by environmental as well as monetary considerations. It is, however, advisable to have an alternative source of power to cope with long periods of little or no sun.

Solar rebate legislation continually varies. In 2020 the upper limit (for grid-connect systems) was a typical 6.6 kW/h. That is also the maximum acceptable to many electricity suppliers for systems that feed excess solar powered energy into the grid.

Figure 2.25. This 11 kW system provides power for Yarrie Homestead (Marble Bar, WA). The entire output is handled by four Outback Power 60 amps regulators. Pic: Peter Wright.

Solar regulators

The system's solar regulator ensures that the output from solar modules is compatible with the rest of the system. The more basic regulators are not able to do this efficiently: a power mismatch precludes them accessing some of the otherwise available energy. This mismatch is due to nominally 12–volt modules producing peak power at about 17-volts, but batteries requiring charging from 13.2-14.7-volts. That energy between 17-volts and 13.2-14.7-volts cannot be utilised by basic solar regulators.

Some of the otherwise inaccessible energy can, however, be made accessible by using MPPT (Multiple Power Point Tracking) technologies (Chapter 16). This technique, used in most high-quality solar system regulators, is well-worthwhile where the solar array is distant from the batteries. It enables the solar array to run at higher voltage, yet still charge at 12, 24, or 48-volts, permitting smaller (and cheaper) cable.

Almost all solar regulators are limited by current - not wattage. A 60 amp regulator can thus handle 720-watts at 12-volts, 1440-watts at 24-volts, or 2880-watts at 48-volts. The higher the voltage at which solar regulators can be run, the cheaper they are per watt. Some MPPT such units accept 300-volts plus.

The size regulator required is the solar array's maximum current in amps, plus about 20%. That extra caters for sunny days with low scattered cloud that momentarily increases solar output. As noted previously, this occurs briefly in areas near a large expanse of water or light coloured sand. In such areas, sunlight is reflected upward from the water or sand, then back down from the underside of scattered white clouds. The modules thus receive both direct and reflected solar input.

Example of solar regulator sizing

A 1500-watt solar array running at a nominal 48-volts generates 31.25 amps. The nearest-size solar regulator available is likely to be 40 to 50 amps. While that is larger than needed, it is good to have some excess capacity. It enables rare excess input to be captured. Do not be concerned, however, if the unit is not quite large enough. That rare excess is blocked (and thus lost), but it does not damage the regulator.

The solar array voltage can be as high as the solar regulator allows - but that part of the system must be installed and connected by a licensed electrician.

Battery capacity

Battery capacity is mostly needed during the evening and early morning. While supplying this, a lead-acid or AGM battery bank is best not discharged by more than 30% or so and be fully recharged by the following midday. Depending on the daytime draw and solar available, this comfortably copes with a few days of zero sun, or a week or so of partial sun without overly discharging such batteries.

Until recently, for most homes and properties, conventional lead-acid or AGM battery capacity needed to be six to ten times the average night-time draw. Solar capacity costs, however, have fallen dramatically. Battery costs have risen. It now pays to have far more solar capacity and less battery capacity. The extra solar provides at least *some* input during long overcast periods: it rarely drops to zero. Avoid having excess battery capacity. If solar energy is not there - it is not there to store, and associated batteries will never fully charge.

Figure 3.25. One of the two 1800 amp-hour battery banks at Yarrie Homestead. Pic: Peter Wright.

Battery chargers

The battery charger must be suitable for charging the type of battery used, and large enough to bring a deeply discharged battery bank up to full charge within 10 or so hours. No high-quality, high capacity charger is cheap. This task, however, necessitates top quality or you may save a few hundred dollars and wreck a multi-thousand dollar battery bank.

Traditional battery chargers (that have transformers) are ultra-reliable, but bulky and heavy and only 60%–70% efficient. The more recent switch-mode (transformer-less) chargers are a fraction of the weight and, at a typical 90% plus, are more energy-efficient.

To assess the charging capacity required, divide the battery bank's capacity by ten. A 500 Ah battery thus requires a 50 amp or so charger. The likely cost is 15%–20% of the price of the batteries.

Generators for battery charging

A generator's minimum size may be determined by various factors but is usually related to that required by the battery charger, plus any other loads simultaneously powered. It is necessary to allow for the extra current drawn by electric motors and conventional battery chargers for the first second or two while starting. This limitation mainly affects petrol-engined generators smaller than 3 kW to 4 kW. Those above that size (and most diesel units) can usually cope.

Big mobile air compressors are a particular case, many of them start-up on full load and need up to 10 times their running current to do so. Their most common outback usage is for blowing out silted bores. Compressors capable of doing that draw a great deal of power. Most people rent self-powered units for the few hours required - rather than scaling the system to supply the rare need. The same applies to big arc welders and similar items.

Conventional battery chargers are typically only 65%–70% efficient. If you have this type of charger, the generator's kVA rating needs to be at least twice that of the battery charger's output in watts.

With the far more efficient switch-mode battery chargers mentioned above, the generator needs to have a continuous rated output of only 40% or so more than the switch-mode charger's output in watts. As noted previously, advise the vendor that the switch-mode charger is to be generator-driven. The reason for doing this is that such chargers prematurely failed and warranty was refused due to using generator input.

Check how close the proposed generator output is to the home or property's average running load. It is possible to provide 100% backup in the event of major solar failure. It is, however, cheaper to postpone running high current appliances and to scale the generator to supply limited domestic power while also able to recharge the battery bank - albeit at a slower rate.

Fuel cells

Eventually, fuel cell technology (Chapter 22) will replace generators. The methanol-fuelled EFOY units (used extensively in boats and RVs) need a battery (ideally LiFePO4) to provide for peak loads.

Chapter 26

Meters & measuring

For solar work, most measurements are of direct current (DC) and voltage (V).

A multimeter is essential, but cheap ones are not suitable. Most electricians use those made by Fluke. These meters are not cheap but last a lifetime.

Figure 1.26. The Fluke multimeter - not cheap but ultra-rugged and reliable. Pic: Fluke.

For any solar array that outputs above 10 amps, a so-called clamp meter (Figure 2.26) is also desirable.

The reason for two meters is that a typical multimeter's current range is only 0 to 10 amps. That range is adequate for checking single 12–volt solar modules of 100 or so watts. Doing so, however, usually requires moving its positive probe to a 'current' socket on the meter. If you forget to move it back again, making a high voltage measurement may wreck the meter, or at least blow a protective internal fuse.

Many solar arrays produce 50 amps upward. Battery current may exceed 200 amps when powering big inverters. Measuring such current requires that clamp meter. These meters measure the strength of the magnetic field created by *one* conductor of a current-carrying cable. The measurement is made by inserting that *single* conductor between the meter's jaws.

Digital multimeters are typically accurate to plus/minus 1%. Clamp meters here are typically accurate to plus/minus 3% or so.

Such measurements can be done far more accurately via a current shunt (Figure 3.26) but are feasible only with permanent installations.

How to measure solar current output

Solar modules are so-called (semi) constant-current sources. Unlike most power sources (such as batteries) they are not damaged by connecting their terminals together. A solar module's current output is measured via a multimeter's 'amp' range if less than 10 amps). If above 10 amps, the solar output is measured by connecting a metre or so heavy cable (a so-called 'short circuit') directly across its output terminals and using a clamp ammeter to measure current flow in that cable.

Doing any of the above requires care. Making or breaking even a 12–volt high-current DC circuit causes arcing. It may burn the end off a screwdriver, or weld a meter probe to a cable.

The safe way is to cover the solar module to preclude output, temporally remove any existing cabling then set up the short circuit. Uncover the module to take the measurement. Then cover it again to reconnect the original cable.

How to measure battery voltage

Measuring deep cycle battery voltage is meaningless unless the battery has rested off-load for at least 12 hours, ideally 24 hours. It needs that time for any charge or discharge to even out.

Because of this a close to fully discharged deep cycle battery may appear to be well charged after only an hour or so of heavy charging. Conversely, a close to fully charged battery may appear to be close to flat immediately after removing a heavy load. Not realising this results in good batteries being discarded - and those useless being retained - while the owner seeks fruitlessly for suspected (but non-existent) electrical faults.

How to measure battery and other heavy currents

A clamp meter (Figure 2.26) can measure starter motor level current (500 amps plus) but with a typically 3% plus/minus error. While that is acceptable for many purposes, others require higher accuracy. This can be obtained by using a so-called current shunt described below.

Figure 2.26. Clamp meter measures current by having one (never both) of the conductors within its jaws. Pic: Digitech.

Current shunts

A current shunt enables virtually any level of current to be measured indirectly but accurately. The concept is basic but ingenious. As shown in Figure 3.26, a current shunt is two thin copper strips. The strip's resistance to current flowing through them is tiny. It does not interfere with that being measured. That resistance, however, introduces an equally tiny voltage (across the shunt) that is proportional to the current flowing and shown on a voltmeter as 'amps'.

Figure 3.26. A current shunt connects in series within a main battery cable feed (positive or negative). Pic: author.

In many vehicles, one conductor of the main 12−volt feed from the alternator and battery to the chassis (and where else needed) is via a single heavy cable. That cable doubles as the vehicle's current shunt – and must not be altered. Most recently made vehicles have that current shunt attached to (or part of) the main battery terminal. It is usually coloured red.

Chapter 27

Installing – legal

Until 2018, it was both routine and legal for anyone (in Australia) who knew how to do so - to install electrical systems that did not exceed 50-volts AC or 120-volts (ripple-free) DC. A new Standard (AS 5139) has abolished that previously. Non-electricians are now restricted to working on (or installing) anything that is over 60-volts DC and 35.4-volts AC.

While this limit does not affect 12, 24, or 48–volt batteries, small solar systems often include a solar array with a higher voltage, connected to the battery via a charge controller using Maximum Power Point Tracking (MPPT) technology.

In such a system, 12–volt solar modules might be connected in series such that the cable leading to the solar regulator carries up to 90-volts. Since this voltage exceeds the new Standard (AS 5139.27), it can only be legally installed by a licensed electrician.

What should a law-abiding DIY person do?

To ensure compliance with the new standard, a shed-mounted existing solar system above 60-volts should not be installed or worked on by a non-electrician. If work must be self-done, options include

contacting an electrician whenever work is required, reducing solar voltages to below 60-volts or limiting battery capacity to below 1 kilowatt-hour of energy storage. For reasons unclear, the new standard applies only to *fixed* installations. A moveable shed appears to be exempt.

Few countries have restrictions as rigid as Australia's. Readers are advised to check the current situation for themselves.

Working on batteries

A human body is a poorly insulated skin bag containing electrolytes and a network that carries oxygen, dissolved in the blood, to the muscles and the brain. That network conducts electricity only too well.

The risk is not voltage per se, but the amount of current that flows and how and where it flows.

Current across one hand's fingers is relatively harmless. Even minor current passing through the brain or heart, however, can be fatal, especially if sustained. Wet skin and especially minor hand wounds significantly increases current flow, and the attendant risk.

Alternating current is more dangerous than DC. It causes severe muscle contractions and the heart to quiver. When dry you may be marginally aware of 12-volts and get a slight tingle from 24-volts, a strong one from 48-volts and a severe jolt from 60-volts DC. Above that, there is a risk of harm, including being thrown off balance from a height. The maximum you may now legally work with (60-volts DC) is borderline. When working on a solar array, first cover the modules with tied-on blankets.

A significant risk when working on batteries (and systems at battery voltage) is of severe burning or battery explosion when fitting interconnecting cables, or tightening terminals on large battery banks. A 1000 Ah battery bank's short-term released energy may exceed a million watts - instantly vapourising a spanner dropped across the main battery terminals. It is worth insulating the handles of such spanners to ensure this cannot happen.

Figure 1.27. One would not have wished to be near this when it exploded! Pic: Vinland.com

Never wear a metal ring while working on 12-volts - particularly around batteries. It is only too easy for one side to contact a live cable while the other side of the ring is touching earthed metal.

Wear overalls and goggles when working on batteries, sealed or otherwise. There is also a risk of explosion if you create a spark near batteries while they are gassing, or accidentally short them out. Smoking near them is lunacy. If you get electrolyte in your eyes, bathe them in clean water and seek medical help.

Inverter safety

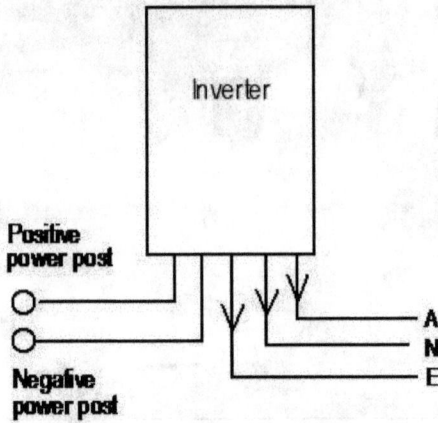

Figure 2.27. Inverter wiring. The cabling on the right is licensed electrician territory.

An inverter's output is just as lethal as from the 230–volt grid. Be aware that some small inverters revert to sleep mode when off-load. They only *appear* to be electrically dead when checked by a multi-meter because that meter does not draw enough current to bring the inverter back to life. Your fingers, however, will. When working on an inverter-driven system, first disconnect the battery positive and negative terminals.

Chapter 28

Installing the system

As with water flow through pipes, electric current dislikes being pushed through cables. A copper conductor's resistance to that flow causes some current to be lost as heat. As a result, the voltage at the load end is always lower than at the source. This loss is called 'voltage drop'.

The amount of voltage drop depends on the conductor's size and length, and the amount of current flowing through it. The smaller the conductor (the copper part that carries the current), the longer the distance (and the higher the current), the higher the voltage drop.

When calculating cable size, take into account the *total* conductor length, i.e. there and back. Ten metres between power source and appliance necessitates twenty metres of the conductor – that can be as two separate conductors or twin conductors within one sheath.

Reducing voltage drop

The effect of voltage drop is an overall loss of energy (in the form of heat). Voltage drop is minimised by reducing cable length, using a thicker cable, or both. Halving cable length halves its resistance, thus halving its voltage drop. Doubling the cable size does likewise.

As the power required to perform work (watts) is a product of current *and* voltage, another way of reducing voltage drop is to use a higher voltage.

A 12-volt electric jug, for example, is rated (say) as 300-watts. As volts times amps equal watts, that jug thus draws 300-watts divided by 12-volts – i.e. 25 amps. If that jug is designed to run at 24-volts, it draws 12.5 amps. If designed to run at 230-volts, it draws about 1.3 amps.

The three variants heat the same amount of water in the same length of time. All draw the same amount of energy and do the same amount of work. Each, however, draws a different amount of current.

For appliance wiring, a realistic voltage drop is about 2%. For a 12-volt jug that is 0.24-volt, but for a 24-volt jug, it is 0.48-volt. The acceptable voltage drop as a percentage remains the same, but in volts, it is doubled. Meanwhile, the current is halved: the acceptable cable size thus becomes a quarter, not half, of that previously. Were there such a thing as a 48-

volt (300-watts) jug, the cable could be sixteen times smaller than at 12-volts!

The above confuse almost everyone at first but think about it for a while, and it becomes clear.

Using higher voltage (where feasible) enables the use of a smaller cable. It is why long-distance power lines run at hundreds of thousands of volts, and some at over a million volts.

Copper is expensive and is becoming more so all the time. By running at very high voltage, the cost-saving on a long high-power run is enormous.

Figure 1.28. Relatively small cables such as this may run at 500,000-volts. Pic: Global Power Line Academy, USA.

All 12/24-volts

It is *feasible* to have a basic system where everything runs at 12-volts. It is cheap, reliable and straightforward. Installation is easy, but heavy cabling is needed to overcome voltage drop over a distance of more than a few metres. Using 24-volts eases cabling but, apart from fridges and water pumps, associated 24-volt products are rare. Such voltages are worth considering, but using an inverter to obtain 230-volts AC allows the use of more readily available and cheaper lighting and appliances.

The inverter decides

In any system involving an inverter, the voltage and size of the battery bank and inverter are determined by the amount of 230-volt power you need. At 12-volts, 1500-watts is about the limit, at 24-volts it is a typical 3000-watts, and at 48-volts it is typically 6000-watts. In each case, the current draw is 125 amps.

Solar array voltage

As previously noted, voltage drop is an issue where the solar array has to be at some distance from the batteries. Most of the now increasingly used MPPT solar regulators accept any voltage from 12 to 78-volts and some hundreds of volts. Because of this, home and property systems tend to have the solar modules interconnected to provide whatever combinations of amps and volts are electrically and legally feasible.

Eight 12-volt modules - each of 10 amps - may be connected to provide 12-volts at 80 amps, 24-volts at 40 amps, or 48-volts at 20 amps. In each case, the wattage (power output) remains the same: at 960-watts.

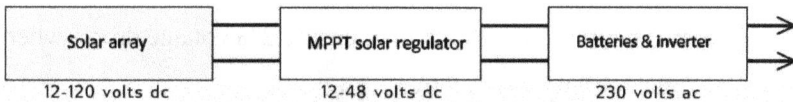

Solar array	MPPT solar regulator	Batteries & inverter
12-120 volts dc	12-48 volts dc	230 volts ac

Figure 2.28. Different parts of the system can be set up to run their optimum working voltage. Pic: rvbooks.com.

Most stand-alone system MPPT solar regulators accept any such voltage - and still charge 12, 24, or 48-volt batteries.

Where the solar array is within a few metres of the solar regulator and batteries, it makes sense to connect the solar modules to run at the same nominal voltage as the batteries.

If, however, the battery bank is at a distance, it makes sense to connect the solar modules such that they run at a higher voltage - and using a solar regulator that allows this. Many do.

Establishing cable size

Requirements for 230-volt cabling are specified in many national standards. That cable's selection is thus specified by the electrician doing the work.

For lower voltage cabling, keep voltage drop to no more than 2% - from beginning to end of any circuit. Some installers cut costs and opt for as high as 5%, resulting in energy being lost (as heat) for the life of the system and appliances not working as well as they could.

Be prepared to pay more for heavier cable. Apart from limitations, such as a conductor being too big to fit into a terminal, you cannot overdo cable size.

For Extra-low voltage DC circuits, establishing conductor sizes traditionally involved following wire table recommendations. These tables list various cable lengths and current flows but typically for a 5% voltage drop, and that is too high for solar installations.

Wire tables are not suitable for work such as this, not least as they introduce gross errors. The method shown below is adapted from the International Standard ISO 10133. It shows the cross-sectional area cable required for any combination of length, voltage and current that results in a voltage drop of 2%. Choose the closest larger size that is commercially available.

By adjusting the (here 0.82) constant, it can be readily adapted for other voltage drops. For 3% use 0.6.15 and for a (a not-recommended) 4% use 0.41.

For conductor size required (in mm²) to ensure 2% voltage drop - where:

L = total conductor length (in metres)

I = maximum current (in amps)

Vo = operating voltage

S = (L x I x 0.82) divided by Vo.

Example:

Solar modules 10 metres from the battery bank producing 50 amps at (say) 48-volts.

L = 20, I = 50, Vo = 48. Thus (20 x 50 x 0.82) divided by 48 = 17 mm². Here 16 mm² is close enough, but AWG 4 is better (if available).

For 3% drop (20 x 50 x 0.615) divided by 48 = 12.8 mm², so 13 mm² is perfect.

Be wary of all voltage drop tables, they are only approximate, and most assume a 5% voltage drop. Use only the method shown above.

How most cable is rated

The current a cable can carry relates directly to the cross-sectional area of its conductor, so electrical equipment makers and the industry specify associated cable accordingly (in mm².) It is simple, obvious and practical and eminently logical. Much of the world uses the International Standards Organisation (ISO) sized cables. The USA and Canada use the American Wire Gauge (AWG) and virtually identical Brown & Sharpe (B&S) ratings that are related to the conductor cross-sectional area. The table below shows how they relate to ISO. Auto cable, however, is a notable and misleading exception that causes significant problems for those unaware of it.

Auto cable

As noted above, ISO cable is rated logically by the cross-sectional area of the conductor that carries the current - and AWG/B&S of any given size has a known and revealed conductor area.

Auto cable (that you are *invariably* sold in chain stores if you ask for 12/24-volt cable) uses ISO numbers for the 1.5mm² to 6.0 mm² sizes mostly used. But auto cable *sizing*, however, is of the cable's diameter *including its insulation*.

Furthermore, auto cable rating takes no account of conductor size: the product is rated by its overall diameter, including its insulation. That insulation thickness varies from maker to maker and even type to type from the same maker. In effect auto-cable is rated by the size hole it can be pushed through.

Compounding the problem is that the cable sizes mostly used for RV and home solar wiring are those most affected: ISO sizes 2.5 mm and 4.0 mm. These may have conductors of less than half the area of the ISO size. Some are even smaller. Even if you emphasise you need 4.0 *square mm in a chain store,* it is all but certain the cable you are sold is under 2.0 mm². Some 2.5 mm auto cable is less than 1.0 mm².

Table 1.28 (below) includes four auto cable sizes, but (except for 6 mm) the comparison with ISO and AWG/B&S is only approximate. Six-mm auto cable (typically 4.59 mm²) is usually a good substitute for 4.0 mm² ISO cable. Some eight mm auto cable is likely to be about 8 mm², but you cannot rely upon that.

When selecting a cable size, it is best to use the formula shown above (under Establishing cable size). A few appliance manufacturers recommended cable sizes, but it is usually the minimum feasible. Commercial installers almost always use that minimum (most assume a 5% voltage drop). The result is a system that loses about 3% of all energy for life - to save a probable few dollars.

Current ratings

A yet *further* cause of cable confusion is the wide-spread practice of marketing auto cable in amp ratings, e.g. '25, 35, or 50 amp'. A 'current rating' indicates the maximum current the cable can carry before its insulation melts. It cannot indicate voltage drop as that relates directly to cable *length*.

Figure 3.28. Overloaded cable and (in this case did) set on fire.
Pic: rvbooks.com

A so-called '25 amp cable' may thus be perfectly fine over a metre or two but has a horrendous voltage drop if carrying that current over 20 metres. Yet go into any auto parts store, and you may hear buyers asking for (say) 25 amp cable - yet never asked by the vendor - 'for what purpose or length?'.

There is nothing wrong with the *higher quality* auto cable as such. Its mm² size is often shown in tiny print on the label or side of the drum, and usually in the store's technical data books. You need to be persistent. Salespeople in such stores rarely know there even *is* a rating difference - let alone that it is vitally important.

Major misunderstanding relating to cable sizes is rife in many homes and RV wiring. It is a known cause of many 12/24-volt DC fridge and other problems. It introduces life-long energy loss in solar electrical systems.

ISO	0.75	1.0		1.5		2.5		4.0		6.0		10		16		25	35		50	70
AWG	18	17	16	15	14	13	12	11	10	9	8	7	6	5	4	3	2	1	0	2/0
Auto		3			4			5			6			8						

Table 1.28. This table shows closest relationships between ISO, AWG and B&S cable sizes. The relationship between those and auto cable is a rough approximation that relates only to up-market auto-cable. The exact (mm²) size of auto cable is, however, usually shown in the small print on the drum and in the makers' specifications.

Cable (voltage separation)

Cables carrying grid voltage (e.g. 230-volts) and low voltage (e.g. 12-volts) must be physically separated unless the latter also uses grid voltage cabling.

It is good practice to keep them separated. Low voltage cables must, in many circumstances, also be protected by conduit.

Specific rules govern the burying of such conduit. Doing so is licensed electricians' territory, but you can save money by digging the trenches. An electrician will advise the required depth. Orange conduit is used for buried cable, grey (or orange) for non-buried cable. Grey conduit can be used for Extra-low voltage cabling. Use an ample size: 20 mm and 25 mm meet most needs, but 32 mm or larger may be required for battery cables.

To pull cables through conduit, tie a rag to a strong line (making sure far end of that line) is well secured, then pull that line through the conduit via suction from a vacuum cleaner. Next, attach the cable to the far end using insulation tape over the first 100 mm or so, and pull it through manually (with another person easing it in at the far end). If necessary, a sticking cable can be eased with soapy water or WD–40.

Terminating cables

Cable terminations usually depend on whatever is being connected. For example, solar modules usually allow only for the conductors to be located under specially shaped washers and secured by set screws.

Connector boxes have holes in which the conductors are clamped under pressure by set screw/s. These boxes are handy for joining multiple cables. The mains voltage versions are readily available from electrical wholesalers and do the job correctly. There is a wide range of these connectors. Some accept cables up to 150 sq mm.

Figure 4.28. Typical connector box.
Pic: solarbooks.com.au

Power posts

To avoid a Xmas tree of cables connected to the battery bank, consider using intermediate power posts, as shown in Figure 5.28. The main heavy current cables (solar regulator, inverter and battery charger) are connected to the shunt positive, and battery negative via such posts.

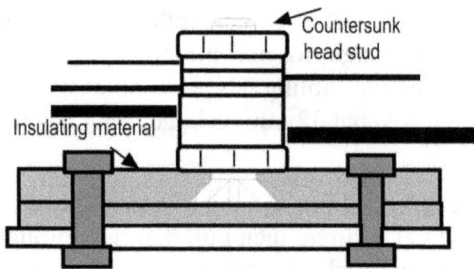

Countersunk head stud

Insulating material

Figure 5.28: This power post was made by the author from workshop scrap. Commercial ones are similar.
Pic: solarbooks.com.au.

Power posts are stocked by marine electrical suppliers and many electrical wholesalers but can be readily made from scrap material. Locate heavy current cables at the bottom of the stack and progressively upwards.

The main cables from the batteries to the main switch, and those that run from that switch to the solar regulator, inverter and charger, must be rated for the maximum current flow. There should not be more than 2% voltage drop from end-to-end.

Crimping tool & lugs

Some devices require crimp lugs. If crimped correctly, lug and cable form a cold weld. If not crimped correctly, the connection eventually corrodes and fails.

Figure 6.28. Crimp lugs are colour coded for conductor size.
Pic: solarbooks.com.au

It is essential to use the correct size lug. If too large, the conductor is trapped at one side rather than being fully crimped. High-quality crimp lugs are sold by electrical wholesalers.

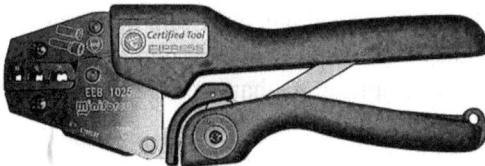

Figure 7.28. Ratchet crimping tool pressure forms a cold weld. Pic: solarbooks.com.au.

Crimping *cannot be done correctly by pliers*: a specialised crimping tool is essential. The better ones have an adjustable preset ratchet mechanism that ensures correct crimping pressure. Big crimp lugs need crimping by a hydraulic tool. If only a few are needed doing an auto-electrician will do it at a nominal cost.

Cable size (mm²)	Lug colour
0.5-1.5	Blue
1.5-2.5	Red
2.5-6	Yellow

Table 2.28. Using the correct size lug is vital.

Protecting cabling

With Extra-low voltage, the main risk is that of fire caused by excess current heating up cabling until its insulation melts. If exposed conductors then touch, full battery current may flow, heating the conductors until they too melt or ignite flammable material. To protect against this, insert a circuit breaker or a fuse close to the source of power. Whichever is used should be rated such that it cuts current flow before the cable heats to a dangerous level.

In practice, circuit breakers inserted close to the power source are best used to protect cables. Fuses (inserted close to an appliance) are best used to protect that appliance against further damage in the event of failure or overload. Over 100 years of US insurance records show fewer electrically-caused fires where circuit breakers protected the cabling. Just why is not fully understood, but there is a statistically significant difference.

DC circuit breakers

It is good practice to include main DC circuit breakers as close to a battery bank as feasible. Each should be rated at about 150% of the maximum current likely to be drawn.

When a DC circuit is broken, an ultra-hot arc forms between the opening contacts and only extinguished when the gap increases. This arc generates a lot of heat and DC circuit breakers' contacts are designed to cope with this. Some have a spring mechanism that causes them to operate very quickly.

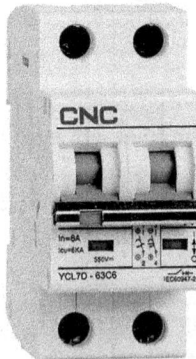

Figure 8.28. CNC's DC circuit breakers are available from 6-63 amps. Pic: CNC.

Several vendors of AC circuit breakers claim their products can be used to protect DC circuits (at up to 48-volts). The units, however, trip differently than with DC. As DC circuit breakers can be obtained at only slightly higher cost, it seems pointless to use AC circuit breakers in DC circuits.

AC circuit breakers

As alternating voltage and current passes through zero voltage and zero current twice on each complete cycle, arcing is not sustained for longer than 100th of a second. Alternating current circuit breakers do not, therefore, encounter the heavy arcing encountered by their DC counterparts. The minimum time for the circuit to be broken following a fault condition is specified. Electricity authorities and regulators increasingly insist that suitably rated circuit breakers be used.

Fuses

Never use tubular glass fuses, nor fuse wire. For Extra-low voltage up to 15 amps, use vehicle blade holders and fuses. Twenty and 30 amp blade fuses are made, but there are increasing reports of fuse holder failures. Mega versions of blade fuses are available and recommended for use from 20 amps upwards, for higher currents use only bolted-in fusible links.

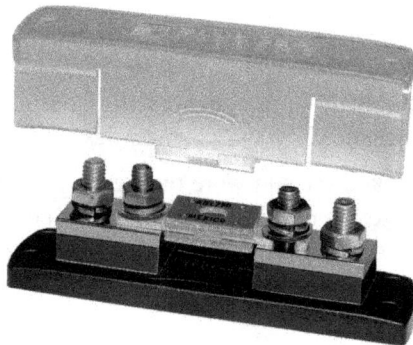

Figure 9.28. This Blue Seas fusible link is available from 35-750 amps. Pic: Blue Seas.

Solar modules need unobstructed sun

Solar modules may be located on a roof or at ground level. Given enough unshadowed space, the latter is preferable, as ease of access for occasional cleaning is more straightforward and safer. To minimise voltage drop the battery enclosure should be as close as feasible to the solar array. Alternatively, run the solar array at a higher voltage, or use a more substantial cable - or both.

Solar mountings

Significant savings can be made by building your module racks. If eligible for a rebate, however, you must discuss and agree on this with your supplier. Not all do – and all have the last say.

Solar module mountings must withstand prevailing winds, especially in cyclone areas.

All except amorphous modules need a 50 mm air space beneath them.

As a generalisation, install modules facing true north and at a tilt angle that is plus/minus about five degrees of your latitude (for non-mobile use). If travelling extensively in an RV, have them horizontal.

Do not be concerned about exactness - errors of five degrees make only a per cent or two difference. If desired, you can optimise input for summer (at the expense of input in winter) or vice versa by building a rack that enables the angle to be changed.

At latitudes numerically lower than 25° (i.e. closer to the equator) there is surprisingly little loss if the modules are horizontal. A slight tilt enables rain to more effectively wash the modules.

Mechanisms that tracked the sun were once common. Solar capacity is now so cheap, however, that it is simpler and cheaper to accept some loss and to increase solar capacity accordingly. Even in numerically higher latitudes, the trend is toward larger fixed arrays.

The module rack shown in Figure 10.28 was made from galvanised roof purlins. Lighter sections can be used where extreme wind strength is not an issue. Cross bracing is essential.

Concrete footings need to be 600 mm by 600 mm in cyclonic areas, but 400 mm by 400 mm or so should be fine otherwise.

Connecting modules

Most solar modules have a connection box on the underside. The boxes have knock-out plastic circles that, accept a 20 mm diameter cable conduit connector. Readily available 20 mm flexible conduit pushes into these connectors (do not use conduit glue as it softens the flex). Connection boxes are often inaccessible once the modules are in place, so finish wiring before bolting them down. To reduce cable length, individually orient modules so that the boxes are close together.

Cabling from the modules can be terminated in a waterproof connector box that also houses the contact breakers, varistors and a common earthing strip. So-called IP 65 and IP 68-rated boxes from electrical wholesalers are fully waterproof if installed correctly. Similar looking but non-rated boxes may lack effective sealing. If there are multiple solar racks, the output from each should be co-joined at the point closest to the batteries. Alternatively, each rack's cable pair can be run to the battery shed and paralleled there. These main array cables may carry substantial current.

Including DC circuit breakers or double pole switches at the solar array end of the connecting cable/s enable you to isolate the solar array's input from the supply cable/s if/when required.

Some solar modules have blocking diodes within the connection boxes. These diodes prevent the battery discharging through the modules at night. As diode blocking is built into solar regulators, these diodes are only needed if no regulator is used, e.g., as with solar water pumping etc.

These diodes drop up to 0.6-volts each. Most installers recommend they be removed. Do not confuse these diodes with the bypass diodes that assist maintaining output during partial shading. These bypass diodes are usually inbuilt within the modules.

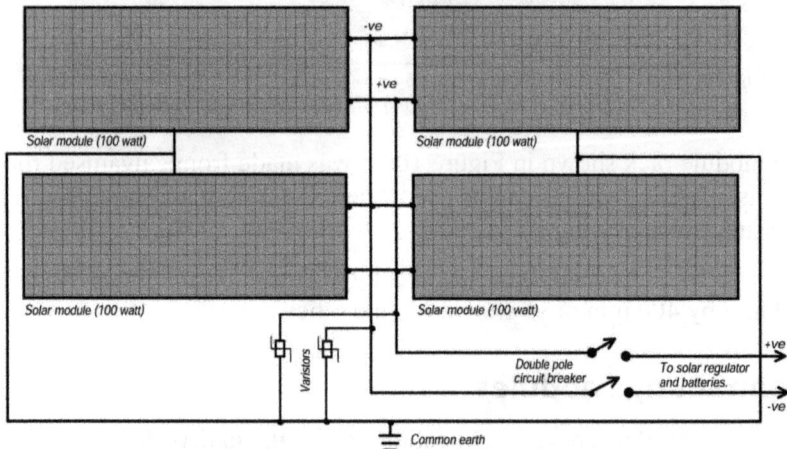

Figure 11.28. This typical small solar array has two pairs of 100-watt modules (each pair is series-connected) with the pairs then paralleled to provide 200-watts at 24-volts (i.e. 480-watts). Note how the two protecting varistors are connected. (Figure 12.28 shows a typical varistor).

Solar modules lightning protection

Lightning protection, via earthing, is not likely to be effective against a *direct* strike. The far more common risk is of nearby strikes inducing high voltages across the supply cables from the solar array to the rest of the system and the solar array's wiring.

Varistors protect against this. They become conductive at voltages of about twice their nominal rating typically in less than 20 nanoseconds, automatically earthing excess voltage.

Figure 12.28. Typical varistor (most are only 5-10 mm diameter).

Use varistors that have a voltage rating of about twice that of the nominal voltage of the solar array. Figure 11.28 shows how they are connected.

In truly static-prone areas, or where lightning strikes are frequent, a varistor can be wired across each solar module. They are readily obtainable from solar suppliers and electrical wholesalers.

Testing series-connected solar modules

Measure the voltage progressively along the module chain (this requires extending meter leads but will not meaningfully affect the result). The voltage should progressively increase. If it does not, then you have probably crossed over positive and negative meter leads on the solar module you are checking. It is only too easy to do.

Excepting for single modules, do not attempt to check *current* unless you have a clamp meter. If you have, check the current with the module's positive and negative terminals strapped together by a short length of heavy wire. Doing so does not damage the modules.

Testing parallel-connected modules

Under a steady sun, and using a clamp meter, cover one module at a time and check the current input to the solar regulator. The current should fall by the same amount for each. If it does not, that module is faulty or not correctly connected, or not fully covered.

If all is well, give the terminals a burst of WD–40 and screw down the connecting box lids.

Earthing solar module frames

Strong dry winds can cause a high voltage static build-up on solar modules and can give you a nasty jolt. That static voltage also attracts dry dust to the module surfaces. You can eliminate this by connecting all metal frames and conductive structural parts of solar arrays to copper-covered earthing spikes (obtainable from electrical wholesalers).

The earthing cables are secured via holes provided for this purpose in the module frame. If the individual modules are well earthed to the metal racks, the earth connection can be taken from that rack. Use at least 6 mm² cables.

Solar regulator

The solar regulator needs to be as close to the battery bank as possible. If, however, the monitoring read-out is part of the regulator (as with most), it is also necessary for that read-out to be easily read. These regulators run fairly hot - ensures ample ventilation.

Most stand-alone energy monitors need to be within 10 metres of the batteries. Some, however, can be up to 100 metres away. They do not need ventilation.

Many solar regulators require either the positive (or negative) solar module output to be connected *directly* to the battery. A light voltage reference cable from that same battery terminal to the solar regulator provides the *exact* battery voltage needed for that regulator to set the charging rate. If that reference cable is excluded, the voltage drop along that regulator to battery cable causes the regulator to 'see' a voltage higher than that of the battery. It reduces the charge rate based on flawed information.

Many solar regulators (notably the Plasmatronic PL series) are wired incorrectly like that. In some instances, it is merely an error. One Australian installer did so for many years to save the few cents cost of a single 1.5 mm² cable. The cost to the owners, however, were systems that never charged efficiently until, and if ever, they were correctly re-wired.

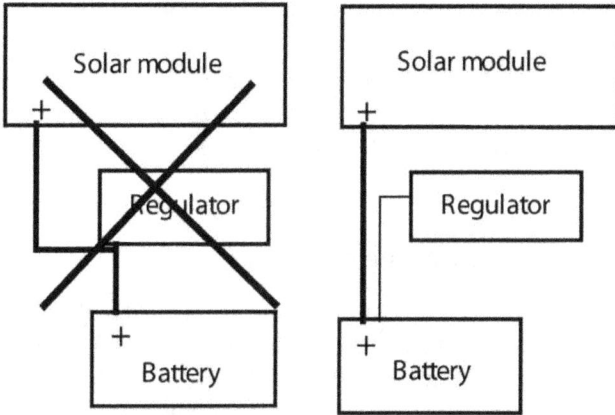

Figure 13.28. Many PL series Plasmatronic regulators and many other type solar regulators are incorrectly installed. They need a direct battery reference voltage (as shown right).

Low-priced solar regulators *have* no provision for a reference voltage. Their battery-to-regulator cables must have no more than 0.1−volt drop in a 12−volt system. Doing so necessitates locating the solar regulator close to the battery, and connected by suitable heavy cable.

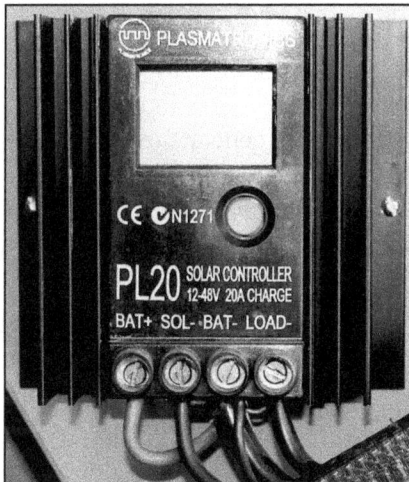

Figure 14.28. A solar regulator's Load terminal can (as shown here) be used to monitor load current − see main text.

Some solar regulators (Figure 14.28) have a Load or similar terminal. If you connect all loads through this terminal, the regulator's readout can show the total current drawn. Such a connection is not always feasible because load current is usually limited by the current rating of the regulator - or to even less.

In many regulators, the load terminal can be used for alternative programmable functions. These may include starting and stopping generators, switching lights on and off at set times, or providing low battery voltage warnings, etc.

Programming solar regulators

Solar regulators need minor programming for the current time of day and the functions mentioned above if they are needed. With some, the solar and battery voltage may not be settable until after the battery is connected. Many regulators, however, accept whatever voltage initially comes in to enable initial setting up.

Locating energy monitors

Monitors usually need to be within a few metres of the associated current shunt. For most, one pair of light signal cables and one pair of very light power cables are run from the current shunt to the energy monitor. Energy monitors usually need programming much as do solar regulators but is usually a simple process.

Locating batteries

Many lead-acid based batteries generate oxygen and hydrogen once their charge rate exceeds 70% or so. The gas ignites (fizzles) at 4% concentration, but explodes violently at around 14%. This gas is generally contained within sealed batteries, but all have safety vents that release gas under excess pressure. While rare, this happens.

Whether the batteries are sealed or not, to safeguard against this, they must be in a dedicated enclosure, shaded from the sun, and vented to atmosphere at top and bottom.

The following minimum enclosure vent area is suggested by the Clean Energy Council (formerly the Sustainable Energy Industry Association). Where: A = minimum area of each vent in square centimetres, C = number of cells in the battery, I = maximum charging rate in amps:
$A = 0.006 \times C \times I$.

Example

A battery room containing four 12-volt batteries (48 cells) and a possible maximum charging rate of 100 amps would require not less than $A = 0.006 \times 48 \times 100 = 28.8$ square centimetres for each vent.

As the emitted gas rises, the upper vent needs locating so that gas cannot be trapped above it. The bottom vent, on the opposite side, should be as low as possible. The cable entry should be at floor level because gas seeps through the tiny gaps between individual conductors in multi-core cable.

Large capacity battery banks

Large capacity battery banks are assembled from multiple smaller units. The bank, shown schematically in Figure XX, is typical of many medium/large systems. Interconnecting cables are needed, and each has a lug at either end. These lugs need to be crimped, *never* soldered. Have them made up by a solar or battery supplier (or an auto-electrician). A hydraulic crimping tool is essential to form the necessary cold weld. These terminals should be protected by adequate insulation.

Battery racks must be ultra-strong and diagonally braced. A single 12−volt, 230 amp−hour battery is likely to weigh about 85 kg. Arrange ready access to battery terminals.

Battery terminal connectors vary in quality, from totally non-acceptable to well-engineered. Avoid cheapies; they are liable to crack.

Battery chargers

Chargers big enough for home and property systems are large and bulky. Transformer versions are massively heavy. Locate the charger close to the battery bank but not in the same enclosure. A battery charger is classified legally as an 'electrical appliance'. As long as it has a supply cord long enough to reach a suitably rated 230−volt socket outlet on a generator you can simply plug it in.

Heavy cabling is needed for the feed to the batteries. Here's an example.

Assume a charger is a 48−volt unit that charges at 70 amps. The charger is five metres from the battery bank, so there are ten metres of conducting cable. The required cable size (S) for a 2% voltage drop is:

$S = (L \times I \times 0.82)$ divided by Vo (here 48-volts).

For example, solar modules 10 metres from the battery bank producing 50 amps at (say) 48-volts.

L = 10, I = 70, Vo = 48. Thus (10 x 70 x 0.82) divided by 48 = 14.5 mm². Here, 16 mm² is perfect.

Current shunts

A current shunt obviates measuring heavy currents directly, and also the need for extended cabling to and from the meter. It also enables current measurement to be cheaper and more flexible.

The current to be monitored is directed to flow through one or more metal strips. Each strip has a small accurately-known resistance. That resistance causes a tiny voltage across it that is directly proportional to the current flow and is displayed on the meter in amps. A shunt adaptor is usually needed.

Most shunts carry 50 amps maximum upwards: you need one that comfortably handles the maximum load or charge current and is compatible with the monitoring device. To ensure compatibility, buy the current shunt and adaptor from the solar regulator vendor. The vendor can advise size and specifications.

The shunt (and adaptor) is usually inserted in the positive cable close to the battery master switch. Some regulators, however, need it to be in the negative lead.

The current monitoring signal is usually taken from the shunt by a pair of light twisted leads. If the read-out is a separate energy monitor, the twisted pair usually connects directly from the current shunt to the energy monitor.

The maximum distance between shunt and monitor is typically 10 metres. Keep the signal cable away from other cables. (Shunts for much longer distances are now available).

If using a shunt with a solar regulator that has inbuilt monitoring, the solar feed cable must go *directly* to the battery, *not* via the shunt. If it goes through the shunt, the solar input is recorded twice.

Switches for 12/24-volts DC

For 12/24-volt DC systems, some switch makers advise that their 230-volt AC switches can be used if limited to 20% of their AC rating: e.g. a 10 amp 230-volt AC switch may be used at up to 2 amps DC.

Figure 15.28. Domestic light switches rated for 10 amps AC can be used for 12/24-volt DC if current does not exceed two amps - just fine for LEDs.

Power outlets for 12/24-volts

Specialised two-pin socket outlets are made for 12/24-volt DC use. They are made by Clipsal (the catalogue numbers are 492/32 plug and 402/32 socket) and are available from some electrical wholesalers and caravan parts suppliers. Avoid cheap cigarette lighter plugs and sockets. Most lack mechanical locking and few safely carry more than a few amps. They tend to arc internally and become a fire hazard. Hella and also Engel make excellent interlocking versions.

Inverters

Inverters may draw 150 amps or more. They need to be close to the battery bank. The transformer-based units are heavy and bulky, but the more recent switch mode units are smaller and much lighter. Most are fan-cooled and noisy. Their noise level should be checked before finalising their location.

The positive inverter terminal connects to the positive system side of the current shunt (or if there is no shunt, the system side of the master switch positive). Inverter negative goes to battery negative.

Cable size is calculated by the usual routine. For example, assume the inverter to be an older 24–volt unit capable of a typically continuous 2400-watts with a short-term surge capability of up to 3600-watts. To cater for some surge capability, allow for 25% or so more than the continuous rating (i.e. 3000-watts or 125 amps).

If that 2400-watt inverter is two metres from the battery bank, (i.e. 4 metres of conductor cable) this is: (4 x 125 x 0.82) divided by Vo (i.e. 24-volts) = 17 mm². Here, 16 mm² fine.

Before switching on the power, triple-check that battery positive *is* connected to inverter positive, and battery negative to inverter negative. If you get this wrong, you may damage or even destroy the inverter.

The 230-volts from the inverter goes to the main switch, circuit breaker/s, residual current detector/s and possibly a solar/generator change-over switch. All must be housed in an approved enclosure.

You can do the mechanical work yourself, but the wiring from the inverter AC output onwards must only be done by a licensed electrician.

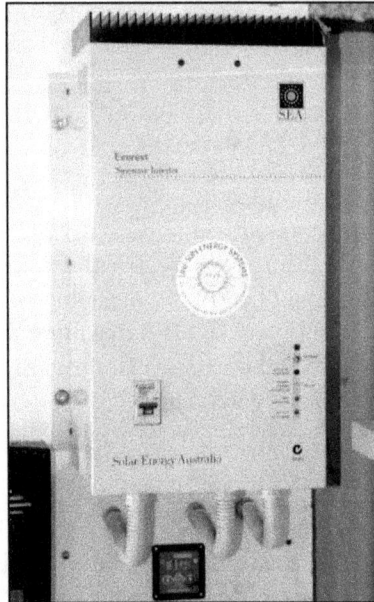

Figure 16.28. This SEA 3.4 kW inverter had a peak output of 11 kW. It was installed in the author's Broome system in 2000 and is still working now (2020).

Generators

Most diesel generators are noisy. Those marketed as 'silenced' are 'less noisy' and may infuriate neighbours. It ispossible to quieten a generator, but acoustic mythology is rife., e.g. the often suggested egg cartons are useless for this purpose.

Most noise from an air-cooled generator is radiated from its cylinder wall and crankcase. It is impossible to quieten this directly, but the sound energy can be mostly contained, and dissipated, by enclosing the generator. The enclosure can be rigid and heavy, such as concrete or limp and heavy, such as sheet lead.

A good compromise is the acoustic foam/bitumen self-adhesive material made for sound insulating engine enclosures in boats. It is sold by marine suppliers such (in Australia) as Whitworth Marine.

As you can experience by driving a car with one window, only a millimetre or two open, sound travels through surprisingly small openings. These can be sealed by sound-insulating tape. You need a sound-absorbing air entry and an exhaust silencer (car units work fine), plus a cooling fan.

Water-cooled generators are easier to quieten, not least because the radiator can be remotely located.

Generators- auto/changeover

The more up-market systems switch a generator on and off automatically under predetermined conditions. For example: 'switch generator on when the battery voltage falls below 46-volts for more than five minutes, then switch back to solar when the battery voltage exceeds 54-volts for five minutes'. The ability to do this is built into up-market solar regulators. The function is usually accessed via the load terminal, or a terminal dedicated to that purpose. It is also built into most energy monitors.

The required is not that hard to do (but often obscured by techno-babble). Specialised knowledge is needed to install the change-over switching, cabling and contactors. Have this work done by the equipment supplier or by an experienced electrician.

Water pumps

Extra-low voltage pumps are often used in RVs and small cabins. Most are 12-volts DC and draw 4 to 7 amps while running and about twice that while starting. Their cabling should be specified for the starting current.

Many pumps are held 'off' by their maintaining pressure. If pressure drops because of a leaking or broken pipe, the pressure switch keeps the pump running non-stop until the supply of water or power runs out. Include an accessible switch so you can turn the pump power off when leaving the premises for any length of time.

These pumps may seize if not used regularly. If that happens, they draw excess current and may burn out. To avoid the latter at least, include a blade fuse at the pump end of the cable.

The Shurflo 'Revolution' pump has an internal bypass. This bypass enables it to run at constant pressure, with water drawn as needed: but entailing higher energy draw. For larger systems consider the variable speed pumps described in Chapter 11. These pumps draw only a third or so of the power of conventional pumps. The big ones run on single-phase 230–volt AC - saving running costs and installation.

Pressure accumulators

Pressure accumulators (Chapter 11) reduce short term pressure fluctuations and pump cycling. They are strong vessels of a few litres to 500 or more litres. Their upper part houses a balloon inflated to about 14 kPa (2 psi) below the cut-in pressure. That pressure is typically 140 kPa (20 psi) in systems that have a maximum pump pressure of about 350 kPa (50 psi).

The pressure tank and switch is best teed into the water system within 10-20 metres of the pump, using the same-size pipe as the main feed. Previous editions of this book suggested longer distances were acceptable (and the author has personally used one at 50 metres). One reader, however, took this to extremes (1.5 km) and ran into problems – due to the inevitable time delays.

Installing a large pressure tank is straightforward. It is well worthwhile for many applications where pumps are otherwise continuously cycled on and off as that tank can reduce pump cycling to once or twice a day. Pump and pressure switch life is extended, and power consumption slashed.

In one installation the pump, that had previously cycled on and off several hundred times a day, then ran for only a few minutes once or twice a day. It reduced energy use from 1.3 kWh/day to 0.1 kWh/day.

Air conditioners

Air conditioners suck warm air from a room and cycle it back as cold air. Most have adjustable air directing fins. For cooling (as cold air falls) direct the fins toward the ceiling, this ensures more even cooling. For reverse-cycle units - that can also heat - direct the fins (and thus heated air) downwards.

Most air conditioners are installed professionally, but installers are unlikely to minimise resultant energy draw. While details are beyond the scope of the book, check ways of minimising heat transfer in and out of the cooled area. Also, ensure the outdoor part of a split system is protected against direct sun, and airflow over and around it is not restricted.

It is feasible to run an air conditioner from an inverter. Care is needed if doing so as many draw excess current while starting up. Some air conditioners have a 'soft start' function that reduces that extra energy needed. Check the air conditioner start-up current before buying. If it is not in the specifications, the vendor should be able to advise.

By and large, a transformer-based inverter rated to cope with the air-conditioner's running current should be fine. Air-conditioning is not, however, an application for most switch-mode inverters.

Refrigerators

The performance and energy usage of refrigerators relates closely to their physical installation and how they are used. The cooling and energy draw of those running on 12/24-volts DC is also affected adversely by voltage drop.

The need for appropriate installation of all fridges is gradually penetrating the industry. Nevertheless, it is still rare for installation to be other than 'push into place and plug it in'. Attending to positioning and ventilation, however, assists in cooling and reduces consumption.

The golden rule is to bear in mind how fridges work. They pump heat from where it is not wanted to where it does not matter. Locate the fridge somewhere cool and out of direct sunlight. Heat insulate the wall behind the fridge and, if feasible, shade that wall by a fast-growing bush or tree.

Ventilation too is essential. Some fridges have cooling fins low down at their rear. Cool air must flow upward and through them, ideally via a screened vent, from outdoors. This air needs directing (via baffles) up and through those fins. These baffles can be aluminium or even stiff cardboard held in place by gaffer tape.

Some fridges dissipate the heat via radiation from their side and sometimes top outer skin. These need low-level cool (side and rear) air vents. A 50 mm gap is needed around a fridge, with nothing above to trap rising air. That air must be vented to the outside.

All the illustrations are relevant no matter whether a fridge is in a cabin, home or RV. In all cases, cool air is needed at the base of the fridge and heat must be enabled to rise with minimal obstruction. That should ideally be to outside the area – or at least to escape.

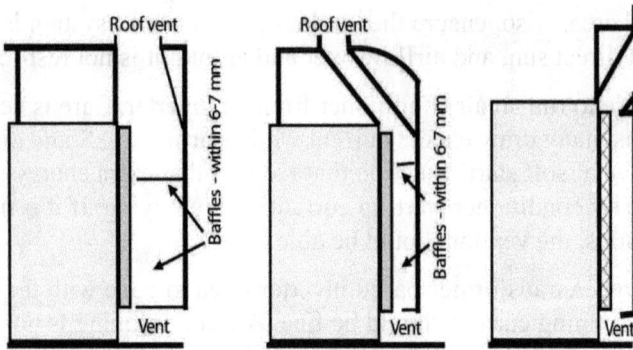

Figure 17.28. **How to make fridges really work**
Left: *baffles direct cool air over the cooling fins, and (compared with the fridge below, where hot air is trapped) the rising hot air has a clear passage to outside.*
Centre: *here again baffles direct cool air over the cooling fins. The rising hot air is directed by a 'chimney' to clear a curving roof line.*
Right: *not ideal but where only a side wall outlet is available for the rising hot air, this will just suffice.*

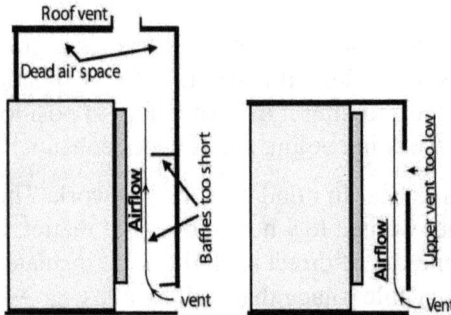

Figure 18.28. **Avoid any or all of this!**
Left: *baffles do not extend in far enough to be truly effective (some air will rise up clear of the fins). The dead space above the fridge is a no no. Hot air builds up and is trapped here, hindering further warm air from escaping.*
Right: *this fridge has no chance of working correctly. Incoming cool air will flow up and out of the upper vent bypassing the fins. Rising hot air is trapped above the vent holding the cooling fins at an elevated temperature. Many RV fridges are fitted like this and none such work properly.*

Fridge cable size

Older fridges run at constant compressor speed. They control their set temperature by cycling on and off - typically for a few minutes each time. Their set temperature thus varies continuously by a degree or so. Many fridges still work this way, but some later units vary their speed as required.

There is rarely a problem with 230–volt fridge wiring, but often with 12–volt fridge wiring. Fridge cooling and energy draw are affected by voltage drop, and many perform poorly because of this.

Assuming a 24-volt fridge drawing 5 amps, 10 metres from the battery bank, an acceptable voltage drop of 2% (0.48 for 24-volts) - and the availability of some leftover 4 mm² cable where:

S = conductor size required (in mm²)

L = total conductor length (in metres)

I = maximum current (in amps)

Vo = operating voltage

S = (L x I x 0.82) divided by Vo

Then, (20 x 5 x 0.82) divided by 24 = 3.42 mm² so that 4 mm² is just fine.

Never compromise on voltage drop with 12/24-volt electric fridges.

Fusing fridges

Some 12/24-volt fridges have internal fuses. Others, such as Engel, have them within the fridge cable plug. If there is not, install a blade type fuse at the fridge-end of the cable.

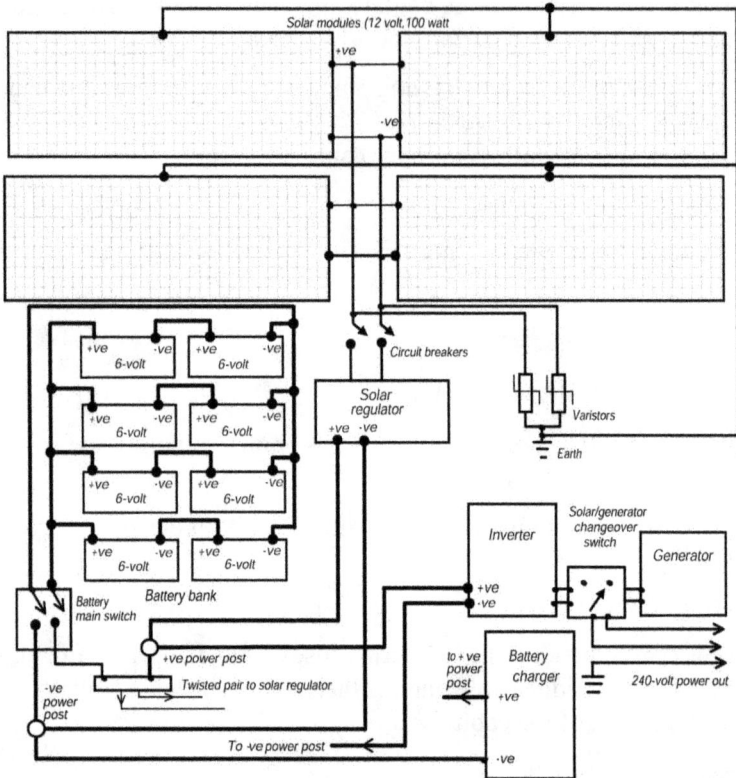

Figure 19.28. Main wiring and components. Four 12-volt modules are paralleled to increase capacity. Four pairs, each of two 6-volt series-connected batteries, are paralleled to provide sizeable current capacity at 12-volts. Regulator/energy monitor wiring vary from maker to maker. A twisted pair from the current shunt may need an adaptor (not shown here) in its feed to the solar regulator. In the interests of clarity, circuit breakers in the battery circuits are not shown.
Pic: solarbooks.com.au

146

Chapter 29

Constructing a Stand-alone System

In Australia, it is currently (2020) legal to design and install stand-alone solar systems as long as the voltage does not exceed 60-volts DC and 35-volts AC. These restrictions effectively limit self-installed solar to RVs and small home and property systems to 60−volt arrays and 24 or 48−volt battery banks.

In all such work, *Solar Success* strongly advises those without electrical experience to obtain competent assistance. The work requires expertise; however, there are many retired people out there willing and able to help.

For any system above cabin-size, consider having an accredited engineer to undertake (or check) the design before you seek quotations for equipment and installation.

If you do self-build, you forgo rebates but often end up in front as installers charge the full retail price for everything involved. Travelling and possibly accommodation adds a great deal more in outback and country areas.

Installation is labour intensive so (at a commonly charged $100 per person/hour) serious savings can be made if you can do part or all of the work. The saving can be substantial for the footings required for ground-located solar arrays. The concrete footings typically involve renting a $175/hour trench digger, then have a concrete truck deliver as little as one cubic metre. Digging trenches, and mixing and pouring concrete is not for the unfit but is not a big task. Hiring a concrete mixer costs only $50 or so a day. (about $50) for the day.

Ground-located solar modules racks cost up to $1000 each (for a four-six solar module unit) plus installation. One quotation for supplying and installing the main array's seven racks, each for six modules, was $8500. These were self-built and installed for less than $1500.

The even-stronger racks were fabricated and installed in one day for under $200 each. The cost would have been $130 or so if cyclonic construction had not been essential.

Subject to minor council requirements, you may build a stand-alone solar system as you want, how you want and where you want. You can locate it on rented land, build it on the back of a flat-bed trailer, have the modules on the roof of a coach and the batteries and inverter in underneath lockers. Or even on a boat towed ashore to power an office! (Figure 1.29).

Figure 1.29: This solar-powered converted A-class catamaran owned by Envirotec's Dick Clarke assists to power his Belrose (Sydney) office when he is not out sailing! Pic: Dick Clarke, Myall Lake, NSW 2013.

The requirements for connecting a 230-volt supply to or from a 'transportable structure' differ from fixed premises and must be done by a licensed electrician. You can, however, legally connect a small portable inverter to a DC supply (most such inverters are 12 or 24-volts). These inverters have one or two 230-volt outlet socket/s inbuilt. Appliances may only be plugged directly into such inverters. They absolutely must not be connected into fixed mains voltage wiring.

Components	3.6 kW	Size	Approx cost
Solar modules	16	175-watts	$3000
Module racks	4	four modules	$2400
Concrete footings	4	400 x 400 x 2000 mm	$200
Solar regulator	1	48-volt, 70 amp	$750
Inverter	1	3800-watts	$3000
Batteries	16	12-volts (230 Ah)	$10,000
Charger	1	48-volt, 60 amp	$700
Circuit breakers	6	100 amp DC	$600
Cable/connectors			$500
Main switch	1	100 amp	$70
Total			approx. $21,220

Table 1.29. The likely cost of a typical 3.6 kW stand-alone outback system, excluding labour.

Table 1.32 shows the likely cost of a typical 3.6 kW stand-alone outback system, excluding labour. The individual prices are those quoted by major local suppliers. Significant discounts apply if most of the components are bought from the same supplier.

Be wary of buying any second-hand system. The chances of finding one that is right for your needs is low. You are likely to scrap most of it.

Chapter 30

Grid-connect

Grid-connect enables excess home and business solar energy to be fed into a grid network, and energy drawn from that network as and when required.

Initially, some electricity suppliers paid solar energy users not only for the electricity fed back but also for that the owners consumed. They recovered the cost by raising electricity prices for all. In effect, non-solar users subsidised solar users by an estimated $14 billion. State authorities later progressively reduced them - but some still extend to 2028. By and large, the grid-connect concept works well.

The grid network

Australia's grid network has electricity supplied by a few mainly coal-fired 'base-line' power stations. One grid supplies eastern states, another supplies the southern part of Western Australia.

These generators have no *immediate* way of varying output (i.e. they have no 'volume control' as such). Because of this, they cannot increase power output to cope with small demand increases: nor can they store any excess.

Electricity usage

Australia's electricity usage peaked between 2006 and 2008, and is progressively falling, primarily due to the increasing efficiency of electrical appliances and lighting, and increasing home and business solar.

Peak demand, however, has not followed that pattern. There is a considerable increase in electricity usage from about 5 pm to 9 pm, with rare peaks of an hour or so on hot summer days when millions returning from work turn on air conditioners. The supply and grid network must cope with such peaks - necessitating it to be much larger than needed most of the time.

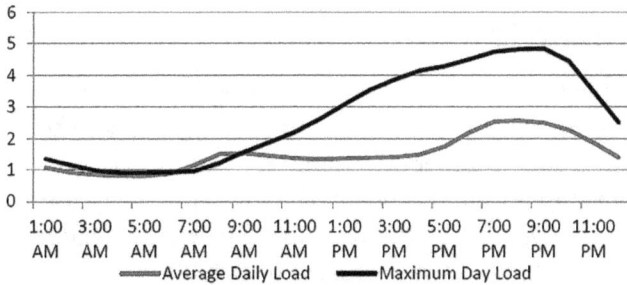

Graph 1.30. This AGL graph (based on Sydney data) shows the difference between the average daily load (dark blue) the highest maximum peak demand that year (black). Pic: AGL.

Our more isolated areas have local grid networks, now increasingly fed by banks of small generators fuelled by diesel, LP or natural gas, enabling individual units to be brought on-line as and when required.

Distributed energy storage

An approach developed and used (some years ago) by the author utilised a solar array, plus units consisting of a solar regulator, a small 230-volt charger with an outlet socket inbuilt, and a 12-volt battery. Controlled by time switches these various size unit charged during off-peak hours and from solar. They supplied localised areas in the home via multi-socket power boards. They could feed into any fixed wiring as the concept is notan electrical *installation.*

Grid-connect

Where solar is viable year-round, about 80% of a typical day's solar input is from 10 am to 2 pm. Peak demand for grid electricity, however, is 3 pm to 9 pm. Most solar input is thus available at times other than when most needed by the grid network and limits feed-in payments.

An increasingly used owner option is to have a 6 kW solar array plus a 12-14 kW lithium battery bank that powers the property during the night. The grid network is retained. Excess solar energy is fed into the grid network (and currently paid for at up to 20 cents per kW/h). Energy is only drawn from the grid when direct, and battery-stored energy is not available.

Going off-grid

Some people (who have grid-power available) nevertheless seek to go off-grid. Its feasibility depends on solar input and whether the local council allows generator usage. It is rarely economic as the battery capacity required is costly (it is far cheaper to use the grid as a 'virtual battery'). Furthermore, home-owners are charged an electrical 'access' charge - whether used or not.

Going off-grid necessitates ample space non-shaded space for the solar array. The average three-bedroom home in temperate areas needs about 7.5 kW - requiring about 75 m² of sun-facing unshaded space. About 30- 45 kWh battery storage typically copes 345-350 days a year. The remaining days rely on a generator and battery charger to recharge the battery bank or supply 230-volts directly.

The cost escalates with increasing reliance on solar. Seeking to provide for 360 days a year without generator power requires doubling solar and battery capacity. Not all urban councils, however, allow you to use a generator.

The above is predicated on current tariff-related issues and those likely to be introduced over the following few years. It varies from state to state. See solarbooks.com.au for updates.

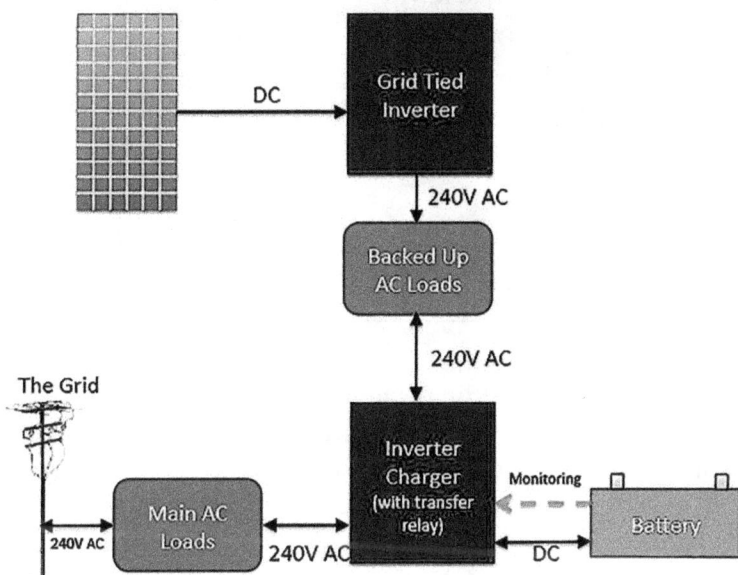

Figure 1.30: Typical AC-coupled system battery-backup system.
Pic: solarquotes.com

Climate-optimised solar

As solar capacity is now so cheap, it is feasible to tilt the solar modules to optimise winter input. Doing, so typically requires them to 20 degrees steeper than your local latitude angle. Consider using that solar input to run high-efficiency 230–volt reverse-cycle air conditioners used as heaters during the day. This works well as the latest reverse-cycle units draw as little as 400-watts at full output.

Strategies to consider now

By following the advice and examples in this book, you can readily reduce energy usage (particularly peak usage). Energy usage can typically be slashed by 30%–50%, sometimes more.

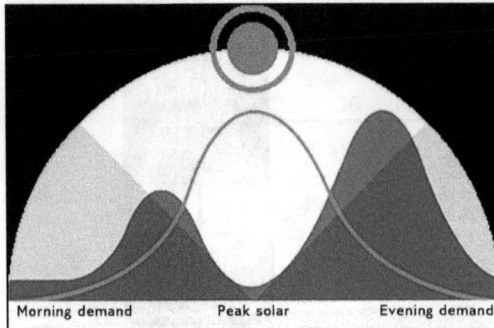

Figure 2.30 - adapted from a pic by Tesla USA.

If your peak demand loading is high and unavoidable, the battery time-shifting approach is well worthwhile. As of 2020, some state governments have started to offer financial incentives to buy Tesla and other brands of back-up batteries – typically of 12-14 kWh.

Chapter 31

Example systems

This system relates to an 80 sq m square cabin in NSW, built in 2002 and progressively updated. It is close to a river with ample water most of the year-round. The cabin is used for weekends, and sometimes longer.

Water was initially hand-pumped with shower, sink and laundry gravity fed from an elevated tank. The toilet was initially whatever needed soil enrichment. A composting unit is now installed.

The original lighting was two outside 100-watt incandescents; an inside 75-watt incandescent and two individually switched 60-watt incandescents for reading. Most were for three to four hours a night.

TV reception was via an old 32 cm (14 inches) unit that drew about 100-watts. There was an even older gas fridge that was replaced by another the same size but marginally younger. All this was run directly, inefficiently and expensively from an old and noisy petrol generator.

Following reading the first edition of this book, the owners then sought to run everything from solar and to install it all themselves. As their priorities were reliability, simplicity, and minimal power usage, they decided to use only 12-volts DC.

The high existing draw was almost entirely through using incandescent globes. Outside lighting was initially replaced a single 25-watt compact fluorescent. The inside light became a single 18-watt compact fluorescent (both 12-volts with inbuilt inverters). The reading lights were changed to 12-volts, 20-watt halogen. The old TV was replaced by a 32 cm (14 inches), 12-volt unit drawing 40-watts.

A Shurflo 12-volt pump was added to fill the water tank automatically when there was excess power. The previous generator-run phone charger was replaced by a 12-volt unit. The total draw was reduced to about 215-watt-hours/day. All ran from two 100-watt solar modules. In 2012, the lighting was replaced by 12-volt LEDs (all of 5-watts) reducing lighting draw to 80 Wh/day.

While two 100-watts modules are retained, the daily draw of a now 135-watt-hours/day would be more than adequately supplied by only one module, even on days of virtually zero sun. The 10 amp regulator and 100 Ah gel cell battery thus cope with total ease.

Winter irradiation is limited by shadowing, but there is still enough input. Minimum irradiation is otherwise about 4 PSH, rising to 6 PSH or more in summer.

Comments

The owners bought all the components from an NSW solar supplier and put it all together over two weekends. It is an ultra-reliable system perfect for a casual weekend lifestyle. Solar input is more than enough, even in mid-winter months when the cabin is rarely used).

Item	Watts	Hours/day	Watt-hours/day
Lighting	395	3.5	1382
TV	100	1.0	100
Phone charger	40	0.3	13
Total			Approx. 1495

Table 1.31. Original energy draw.

Item	Watts draw	Hours/day	Watt-hours/day
Lighting	20	3.5	80
TV	40	1	40
Phone charger	10	0.2	3.3
Water pump	50	0.2	10
Total			Approx. 135

Table 2.31. Present-day current draw

Example 2 – mid-sized cabin

This two-bedroom holiday cabin in Queensland with verandah sleep-outs was built in 1980 for holiday use. In 2008 the owners decided to use it as a more or less permanent home, but have a solar-equipped motor home used for about each year during the winter. They also travel extensively overseas. They sought more facilities, putting their limited funds into travelling rather than their home.

The cabin initially had a basic 230–volt system run from a petrol-driven inverter/generator. There were a small LP gas fridge and water heater. A local electrician suggested water heating be replaced by an LP gas/solar unit, the incandescent globes be replaced by compact fluorescents. The owners installed an 81 cm (32 inch) LCD TV - and to use LP gas for cooking. The owners felt that having been used to living with a 150-litre fridge in their motor home, they'd be happy with a 170-litre, 24–volt unit in the cabin.

The owners had not previously considered grid-connection, but as grid power became available, it made sense to use it. A (then) $8000 rebate would substantially pay for the small 1 kW system required. The (then) 44 cents per kWh feed-in tariff resulted in a typical $45 a month return (when the cabin was not used) and until 2028.

With grid-connect available, it seemed pointless to continue to run a less efficient 24-volt fridge via a 230/24-volt converter. An ultra-efficient, 220-litre, 230-volt unit was installed instead. With the bigger but efficient fridge, half the lights on for about five hours each day, and reduced appliance and contingency allowances, the average daily usage was 4.83 kWh/day.

The grid-connect system installed supplies all the owners' day time needs except in mid-winter when they are rarely there (and during which an average of 3.5 kWh per day gets pumped back into the grid). There's a more substantial surplus in mid-summer. Year-round this system is more grid supplier than grid user.

Updates

A new LED TV uses about 100 Wh/day. A new washing machine uses about 200 Wh on its cold cycle. The lighting is now mostly 5-watt LEDs, and energy usage about 2.25 kWh/day.

Item	Watts drawn	Hours/day	Watt-hours/day
LEDs 5 W	30	5	150
Fridge 220 litre			300
TV 81cm (32 inch) LED	100	2	100
Washing machine	200	1	200
Dishwasher	700	1	700
Kitchen appliances	300	1	300
Contingencies	500	1	500
Total			Approx. 2250

Table 3.31. Current (2018) energy draw.

Example 3 – twenty acres in WA

The property owners are semi-retirees with twin daughters – all have a horse. The house is energy-efficient and cross-ventilation negates any need for air conditioning. The (3.5kW) stand-alone solar system originally had generator assist year around. The owners wished to run mainly on solar but needed far more lighting. The initially high usage is shown in Table 4.31.

Item	Watts drawn	Hours /day	Watt-hours/day
Lighting 10 by 22 W	220	4	880
Lighting 20 by 100 W	2000	4	8000
Fridge 500 litre (old)			3300
TV 66 cm (26 inch) (old)	300	3	900
Washing machine	1350	2	2700
Dishwasher	1400	2	2800
Appliances, etc	500	1	500
Computers (3) – desk-top	900	6	5400
Water pumping	1750	2	3500
Horse stuff	500	2	1000
Power tools, etc	250	1	250
Various	500	1	500
Total			Approx. 29,730

Table 4.31. Initial energy draw

All incandescents were replaced by compact fluorescents and assisted by automatic time switches; average usage was reduced to three hours a day. The fridge was replaced by a new 507-litre fridge-freezer drawing 1600 Wh/day. The oldish computers were replaced by laptops.

Replacing 20 mm diameter piping by 50 mm for the main 400-metre run, and adding a 500-litre pressure tank slashed pumping draw by 75%. The washing machine was replaced by a large top loader that draws 200-watts using cold water.

The above changes cost about $10,000, but reduced energy draw to just under 11 kWh/day. The changes enabled existing solar capacity to provide self-sufficiency apart for occasional generator use during June/July.

Item	Watts drawn	Hours /day	Watt-hours/day
Lighting 30 by 5 W	150	3	450
Lighting 30 by 10 W	300	3	900
Fridge 507 litre)	1600		1600
TV 66 cm (26 inch) new LCD	150	3	450
Washing machine (cold)	200	1	200
Dishwasher	1400	2	2800
Appliances, etc	500	1	500
Computers (3) laptop	280	3	840
Water pumping	500	1	500
Horse stuff	500	2	1000
Power tools, etc	250	1	250
Various	500	1	500
Total			Approx. 10,000

Table 5.31. Energy draw, (2018).

Further update

Subsequent savings were made changing to all lighting 5-10-watt LEDs. A 48–volt high brushless DC water pump reduced water pumping from 1750 W/h to only 500 W/h a day. The total saved (about 1000 Wh/day) would have enabled a smaller solar system to be used (certainly less battery capacity). The existing one now only rarely needs the generator used.

This installation exemplifies this book's central thrust: to minimise use before you think about solar: in this case from close to 30 kW/day to under 10 kW.

Example 4 – an ongoing (Queensland) system

This system has been documented here for close to 20 years. It was initially devised by a then 16-year-old student who progressively improved the system - and still provides updates for this book.

The home is a small property on a largish block in south-east Queensland and is connected to the electricity grid. It originally had un-insulated timber walls and a marginally heat-insulated, dark corrugated iron roof. Cooking was via gas, and water heated by a couple of instantaneous electric heaters dating back to the early '50s. The property was lit throughout each night (for perceived security reasons) by four 100-watt incandescent globes plus three 150-watt spotlights.

Natural lighting was poor and the beige walls absorbed light, requiring indoor lights (then all incandescents) to be on 16 hours a day. The living area had three such lights, of 75-watts each, the kitchen had two, each of 100-watts. The three bedrooms each had a 60-watt reading light, plus a 50-watt overhead light. The windowless bathroom had a 40-watt incandescent globe – left permanently on.

There was (until 2017) a 66 cm (26 inch) colour TV which, back then was left on all day. A VCR was also permanently on as was a relatively modest hi-fi system hardly ever used.

The ineffective fridge was over 25 years old and leaking 'cold' to the extent that it had no 'off' cycle. The energy-hungry dishwasher was run twice a day. Overhead fans in each room attempted to cool the summer heat, and bedroom fans ran all night.

An old 200-litre chest fridge was left on year-round despite being used only to chill beer for an occasional party. An electric fruit dryer was used for days on end, but often to dry less than a kilo or two of fruit. Of the family's little money, a good part went to pay the huge electricity bill.

The daughter, an environmentalist with a scientific bent, quantified the household's electricity usage as her final year school science project. The table shown here is a summary of that data. The project noted that the writer 'was appalled by the energy waste'. Now an academic in this area she left school for university soon after, but not before her and a local electrician making significant changes.

Lighting alone was responsible for 40% of the total (then well over three times the national average). Even though the town had a close to zero crime rate, her father argued that 'we don't get break-ins with lights like this!' - but agreed to cut it back to six 18-watt fluorescents.

She heat-insulated the walls, had the roof painted light grey and the interior in less light-absorbing colours. That, together with curtains open during the day, eliminating the need for daytime lighting. Increased ventilation also eliminated the need to use two of the fans, and halved usage of the remaining four. All inside light globes were replaced by compact fluorescents. The old electric water heater was replaced by a 330-litre solar water heater.

Presenting her father with the knowledge that running the chest fridge all year round was costing $200 (each slab of beer costing far more to cool than initially buy), solved that issue there and then.

The dishwasher's draw was absurd and was replaced without 'too much opposition' by a larger but still old one that was used only once a day. The worn-out fridge was replaced by a more efficient unit of similar size. Minor changes included replacing the bathroom light switch by a motion sensor and using the washing machine on a cold cycle only.

The 'leaving the TV on' culprit was her 11-year-old brother. She told her father how much it costs in electricity and he deducted it from the boy's pocket money.

The costs

Excepting for the electrician changing the bathroom light switch and installing the solar water heater, all work was done by 'a not willing' father and the daughter.

Roof: cleaning and priming, plus two coats of light colour paint – approximately $350

Insulating bedroom walls – approximately $200

Painting inside walls in the living area and kitchen – approximately $75

Twenty 18-watt compact fluorescent globes – approximately $100

Second-hand but efficient fridge – $500

New solar water heating (incl installation) – $3100

Total – approximately $4550.

The savings

Initially, electricity cost $2680 (52.5 kWh day). Despite price increases, the cost of the changes was recovered in two years. Subsequent savings helped pay the cost of a 1.5 kW grid-connect system, implemented when the daughter finished her first degree. She subsequently completed an M.Sc. in electrical engineering. During summer, solar now provides all day-time power.

This example shows just how much energy and money can be saved. It was only done following an environmentally aware teenager's realisation that power usage was seriously wrong.

An interesting aspect is that a then 17-year-old persuaded her family to make the changes. I am grateful for a comment made by one of those kind enough to vet the draft copy of the first edition of this book. 'It's hard to change what you do yourself; it's even *harder* to change what others do'.

Update

The outside lighting was later replaced by four 10-watt LED floodlights and the TV by a new one that draws 70-watts. This reduced daily draw by another 0.5 kWh/day. It was thus possible to cut usage to about 8.80 kWh/day.

My ongoing thanks to the daughter for updating information for these ongoing two pages.

Item	Watts	Hours /day	Watt-hours /day
Lighting – living area	225	18	4050
– kitchen	200	18	3600
– bathroom	40	24	960
– external	850	12	10,200
Dishwasher	2500	3.5	8750
Washing machine	200	1	1200
Fridge (425 litre)	210		5040
Fridge (outside)	150	24	3600
Kitchen appliances	500	1	500
Fans (six by 75 W)	450	8 (min)	5400
TV	395	8 (min)	3160
Various (incl. water heating)	3000	2	6000
Total draw			Approx. 52,460

Table 6.31. The original draw - three times the national average.

Item	Watts	Hours /day	Watt-hours /day
Lighting – living area	54	5	270
– kitchen	54	3	270
– bathroom	18	1	18
– external	108	12	1296
Dishwasher	2000	1.5	3000
Washing machine	200	1	200
Fridge (425 litre)	150		1800
Kitchen appliances	500	1	500
Fans (four by 75 W)	300	4	1200
TV	395	3	1185
Various (incl. water heating)	500	1	500
Total draw			Approx: 10,240

Table 7.31. The above shows the initially-reduced total. The daily draw is now (2020) about 8.8 kW/h per day.

Chapter 32

Living with solar

Stand-alone solar makes one conscious of energy wastage even if there's ample power to spare. Lights are not left on unnecessarily, TVs are not left running with the sound turned down, nor turned off only from the remote control. One buys only energy-efficient appliances. Dripping taps are repaired, and less water used while showering.

Then visitors arrive! Lights are left on day and night and showers used until they run cold. The TV is left on all day with the sound turned down. Computers used to play video games are left on all day and night.

While living in the Kimberley, we partially solved it by having a visitors cabin with its own bore-fed solar hot water system. The cabin also had bore-water for toilet flushing and laundry. Visitors still wasted water, but as we used less than 0.2% of our allocated yearly amount, those 15-minute showers were not a problem.

Significant problems can arise with caretakers unaccustomed to life away from the grid. Energy usage may escalate. Batteries can be wrecked by ongoing flattening. One solution is to have a grossly oversized system or an automatically starting and stopping generator that recharges the batteries - and accepting the resultant fuel bill. Or no caretakers.

Seeming lack of solar input

Systems with adequate solar capacity often surprise new users by there being no extra input during days of full sun. There's usually nothing wrong. The system monitors the battery charge: when 'full' it reverts to a low input 'floating mode'. It has the *potential* to supply more energy but only does so if needed - as when a heavy load is applied.

165

Battery maintenance

The wet batteries that need topping up with distilled water every six weeks or so are still used on some large outback properties.

Sealed batteries have a shorter life span than well maintained wet cell batteries, but unless they *are* it pays to use sealed versions. Monitoring is essential, but once the system is set up correctly, there is little to do except checking daily battery percentage charge

Leaking batteries

People unaccustomed to solar tend to worry about more energy coming in than is used: and suspect batteries somehow 'leaks'. In this universe at least, energy cannot be turned from one form into another with 100% efficiency. Battery loss is minor and varies with battery type, capacity and age.

Solar modules

Solar modules need occasional cleaning. Dry winds cause static charges that attract dust. A light shower then turns that dust into mud - but insufficiently to flush it off.

Wash modules regularly using water and detergent. Rinse with clean water plus a tiny amount of detergent (it has anti-static properties). Use a squeegee to remove surface water and let the modules dry naturally. Do not wipe nor polish them: doing either generates dust-attracting static.

Chapter 33

Our solar systems

Solar usage for us began in early 1994. My wife and I fitted an 80-watt Solarex solar module, regulator and 100 Ah deep cycle battery (powering a 40 litre Engel fridge and few lights) to our restored 1974 Kombi. We drove that Kombi extensively across the outback.

Figure 1.33. Our rare 1974 Westfalia VW Kombi (and the author's wife, Maarit). Pic: rvbooks.com

In 1996 we converted an off-road OKA into a motor home. Two 80-watt Solarex modules charged a 250 amp–hour lead-acid battery bank that ran a 70–litre Autofridge, a suitcase-sized Westinghouse satellite phone, and ten halogen globes. The solar modules also drove a computer, printer and a sat-phone linked fax machine.

Figure 2.33. Our ex-mining coach (a 1994 OKA) crossing the huge Big Red sand-dune in the Simpson desert. Pic: Maarit Rivers.

In 1998, we bought 10 acres of ocean frontage north of Broome. The OKA then became home while we self-built our own. The vehicle also served as cyclone-protection - once withstanding 180 km/h gusts.

The property had (and still has) no supplied services. With the closest grid feed $500,000 away, we designed and built an (initially) 2 kW system that drove a 48-volt 3.8 kW SEA inverter.

Our solar system's 11 kW surge capacity enabled us to self-build our 186 square metre concrete, steel and glass home. We used none but self-generated power.

The Broome house's structure

The house was initially designed by an architect who is no longer practising. His concept was excellent, but it needed significant changes to meet cyclone requirements. The final structure was closer to bridge or ship-building than house construction. No Kimberley builder would touch it! Except for contracting the roofing and concreting, my wife and I built it ourselves with help from a local man who had some building experience.

The house has an ochre-tinted, sealed and polished 200 mm concrete floor. Forty 100 mm by 100 mm (4 mm thick) RHS steel columns at 1.8-metre centres tie down the 4.3-metre double curvature roof.

External 'walls' are cyclone-proof toughened glass with stainless-steel security mesh sliding screens. There are thus 15 'front doors' with that used depending on whichever was climatically most suitable. There was no need, nor provision, for air conditioning. The bolt-together design and construction allowed us to move in within five months.

Before commencing building, the author's Finnish wife (Maarit) obtained Arc, MiG and TiG welding certificates, plus a Production Engineering certificate from Broome TAFE. She did much of the welding on the building's massive rolled steel joist (RSJ) outer frames and was adept at carving up concrete with her 12-inch angle-grinder.

Figure 2.33. Here, Maarit's wielding her 12-inch angle grinder.

Extending the solar

The original 630 Ah (48-volts) battery bank was ruined by chronic overloading during a three-month absence. We replaced it in 2006 when we moved the system closer to the house.

The existing modules, plus a further six (64-watts) and twelve (130-watts), were bolted to cyclone-proof racks, facing north at the area's 18° latitude angle. We connected modules to provide a nominal 72-volts, and an actual 110-volts or so off-load.

Figure 3.33. Our self-designed and built 3.8 kW solar array.
Pic: rvbooks.com

The output was fed via twin DC circuit breakers via 25 mm² cables buried in 32 mm conduit to a cyclone-proof shed 25 metres away. It led from there, via a 75 amp fusible link, to an Outback Power MPPT 70 amp solar regulator to feed 16 by 230 amp–hour Exide gel batteries housed in a specially built battery room accessible via an external cyclone-rated roller shutter door.

The regulator monitored incoming energy only, so we added a Xantrex Energy Monitor. Later expanded to 4 kW, the system generated a year-averaged 18 kWh/day: more than enough to run the property and its extensive irrigation system year around.

We had a 6.0 kVA back-up generator and 50 amp NZ-built Stanbury Engineering charger. Both, however, were needed for only 50 hours - while moving the solar system to its later location.

Figure 4.33. Part of our previous 10-acre property at Coconut Well, Broome. In front of the house is a tidal lagoon that filled twice daily from the Indian Ocean (in the background). Pic: Just before its sale in August 2010. It was a hard place to leave!

Water

The house has deep, wide full-length stainless steel gutters that, for cyclone protection, were located concealed between the roof and the ceiling. These ran, via 100 mm downpipes, to two horizontal 200 mm main pipes - then to a 14,700−litre holding tank, behind and below the house. This tank accepts the Kimberley's massive short-term cyclonic rainfalls, often filling in less than 30 minutes. Water is pumped from the holding tank to interconnected 100,000 litre and 25,000−litre tanks 75 metres away, via a 0.75 kW pump and 50 mm pipe.

Holding about 80% of the average rainfall of 670 mm, the full 125,000 litres was invariably filled during the wet season (in 2008 inside 10 days). Rainwater was used for all household needs, including toilet flushing. There was usually 20% left over each year.

The water was pumped to the house from the holding tanks via twin (linked) 500−litre pressure tanks at an initial 360 kPA (50 psi).

Irrigation

The bore (producing crystal clear water) initially had an on-line Grundfos SQ bore pump, but this proved less than successful. We understood later that this pump is not recommended for use in the Kimberley. We changed to a new system that had a bog-standard 0.75 kW unit bore pump that filled a 21,000–litre tank adjacent to the bore.

The bore pump filled the tank until cut off by a float switch and automatically turned back on when (a) the batteries reached float voltage or, (b) the tank level dropped below 13,600 litres. The former signal was derived via an energy monitor, the latter from a second float switch. See Chapter 11 for details.

A separate 1.13 kW pressure pump fed the irrigation system, swimming pool, and outside taps. This pump was controlled by a pressure switch that cut in at 140 kPa (20 psi) and out at 280 kPA (40 psi). A 500–litre pressure tank aided efficiency. The mainly drip-feed irrigation supplied about 3500 litres/day via five separate valve-controlled circuits.

Gate valves were used in each circuit to reduce pressure to 50 to 70 kPa (7-10 psi) - necessary as tropical sunlight quickly weakens irrigation piping and clips.

Swimming pool

The 31,000–litre pool used bore water. About 2000 litres was renewed each day by running the irrigation feed via the pool. A pipe rose from just above the bottom of the pool, to a little above normal water level, and then dropped back down and into the irrigation system. When a timer valve let in water, the level rose until it reached the horizontal section of pipe and then down four metres into the irrigation system.

The pool's 48–volt brushless DC motor's circulating pump ran 12 hours a day from four dedicated 120-watt modules. As there was no need to circulate pool water at night, no battery storage was required.

The end of an era

Our 12-year sea-change came to an end in late 2010. With considerable reluctance, we sold the property, relocating to a house at Church Point (35 km north of Sydney) directly overlooking Pittwater.

While it is an environmentally built house, the previous power owner's power usage exceeded 30 kWh/day.

We tracked down over 4 kWh/day in twenty or so phantom loads. The most useless was an entry-door chime's power supply that drew 960 Wh/day. Another was an instant water boiling/freezing device. This unit consumed over 2.0 kWh/day without water even being drawn.

Figure 5.33: Solar array on the roof of the author's home in Church Point, Sydney. It faces north-east at less than latitude angle to also capture the sky-haze reflected irradiation from Pittwater directly in front of the elevated house. The peak input from the (now) plus 6 kW (and 14 kW/h Tesla battery) system is over 40 kWh/day during summer.
Pic: solarbooks.com.au.

The house had over seventy 35-watt halogen globes, plus walls lit up by 10 spot-lights - each drawing 150-watts. Those and all external lighting were replaced by LEDs. We removed and sold the boil/freeze unit, and replaced the door chime by a manually rung ship's bell.

We initially a 2.4 kW solar grid-connect system, a 420−litre solar water heater, and a Solar Venti space heating unit (but proved far less successful than hoped).

In 1999 we extended the solar capacity to 6.5 kW and added a 14 kW Tesla battery.

Heating is via reverse-cycle air conditioning powered by the solar array during the day and by the battery for a few hours each night.

Year-round, despite the house being used as the author's office, and his wife's (Maarit) practice as a child psychologist, we have excess solar even during much of the winter. We sell as much as 35 kWh/day back to the grid (at a current 20 cents per kilowatt-hour).

Chapter 34
Technical Terms Explained

Alternating current: A flow of electricity that alternates in polarity. Each double alternation forms one complete cycle (Hz). Alternating current's working action is like a two-person saw that is alternately pulled to and fro, cutting on both strokes. It is abbreviated as AC. (See also DC).

AGM (Absorbed Glass Mat): A sealed lead-acid battery that holds electrolyte within a fibrous silica glass mat.

Amp: An abbreviation of ampere; the unit of measurement for instantaneous electric current. See also 'current'. Amp is abbreviated as A.

amp–hour: This brings a time element into the above. A current of one amp that flows for one hour is thus one amp-hour. It is abbreviated to Ah or A.h. (In the SI system of units, amp-hour is shown as amp-hour).

Available capacity (battery): The charge, in amp–hours, that a battery delivers under specific conditions.

Battery: An assembly of cells that stores electrical energy, in the form of chemical energy, for later use.

Cable size: Unless otherwise defined, cable size generally refers to the cross-sectional area of its copper conductor that carries the current.

Charge rate (of a battery): The rate of current flow (in amps) that restores its available capacity.

Circuit breaker: A switch that automatically protects against excess electric current.

Current: The rate of flow of electricity (akin to the rate of water flow in a pipe). It is expressed in amps.

Current shunt (also shunt): A short length of heavy copper used to monitor high currents remotely. Current flowing through a shunt causes a proportional voltage to be generated across it. That voltage is shown on a meter (in amps). The shunt negates the need for the massive cabling otherwise needed to measure heavy currents directly.

Conductance: The ability of a material to conduct electricity – hence the term 'conductor'.

Cycles per second: Now more commonly expressed as hertz (Hz), i.e. the number of times each second that alternating current completes its bi-directional movement.

DC (direct current): That associated with solar modules and batteries. It is called direct current because it flows only in one direction in action akin to a band-saw. DC current is expressed in amps. The abbreviation is DC.

DC-DC converter: An electronic device that increases or reduces the voltage of direct current.

deep cycle: A term for batteries that withstand ongoing deeper discharging than (say) starter batteries. The term can mislead because deep cycle batteries degrade quickly if *routinely* deeply discharged.

Depth of discharge: The percentage of capacity drawn from a battery, 40% discharge implies 60% remaining.

Diode: A device that allows electrical flow in one direction only (an electronic version of a one-way street).

Earth: The earth (in the USA, the ground). In many countries, the neutral conductor of alternating current equipment is connected to an earthing rod that is pounded into the soil.

Electricity: A form of energy produced in various ways that result in a flow of electrons.

Electron: An elementary particle that has a negative charge. An electric current is a flow of electrons.

Electrolyte: A liquid or semi-liquid that carries electric current by ion movement (rather than electrons).

Feed-in tariff (gross): Payment made by an electricity producer (or other body) to the owner for *all* output from a grid-connect solar system, regardless of whether used within that system or fed into the grid network. (Nowadays rare).

Feed-in tariff (net): Payment made by an electricity producer (or other body) to the owner for the *surplus* output from a grid-connect solar system that is fed into the grid network.

Float charge: The current required to balance that lost by (a battery's) internal self-discharge.

Fuse: A heat sensitive device that melts and cuts off power to protect against the excess flow of current.

Frequency: The number of cycles per second by which an alternating current completes its bi-directional movement – see also 'cycles per second'. Frequency is usually expressed in Hertz – or abbreviated as Hz.

Gassing: The emission of hydrogen and oxygen from the electrolysis of water from a charging battery.

Gel battery: Sealed lead-acid battery in which the electrolyte is held within a silica gel matrix.

Grid: a national or local network that distributes electricity.

Grid-connect: Suitably configured systems that draw power from the grid network or supply power to the grid network.

Hydro-turbine: A rotary device that converts the kinetic energy in a flow of water into electricity.

Insulation: A material that strongly resists electron flow.

Inverter: A unit that converts direct current to alternating current (typically to 230-volts AC).

Irradiance: Direct, diffuse and reflected solar irradiation that falls on a surface. It is expressed in kW/m^2.

ISO: International Standards Organisation.

Joule: Basic ISO unit of energy and work – 1 joule/second equals 1-watt.

Maximum Power Point: The combination of voltage and current where power output peaks.

Maximum Power Point Tracker (MPPT): A technique that, by continually optimising voltage and current, enables a solar regulator to operate at peak efficiency. Also known (in a more straightforward form) as a maximiser.

NOCT (Nominal Operating Cell Temperature): The actual temperature of cells inside a solar module at 25° C ambient is typically 47° C-49° C. The NOCT output allows for this and other losses and is typically about 70% of claimed maximum output. NOCT data more accurately indicate what solar modules put out in real life. The NOCT is usually disclosed on a data panel on the rear of each solar module - or buried in the small print.

Parallel (connections): Connecting negative to negative, and positive to positive of (say) batteries and solar modules. Doing so increases current; the voltage remains unchanged. See also 'Series connection'.

Peak Power Point: That point of the voltage/current output of a solar cell or module where the product of the two results in maximum power.

Peak Sun Hours (PSH): The number of hours per day equivalent to the sun's irradiance at 1 kW/m².

Phantom loads: Many electrical devices continue to draw some current if switched off only at the device. The current drawn is called a 'phantom load'.

Photovoltaic: A substance capable of generating voltage from visible or other radiation.

Power factor: The ratio of the power (in watts) that is consumed in an alternating voltage circuit, to the power (in volt-amps) that is being drawn from the power source.

RCD (Residual Current Device): A safety device that continually checks that current flow in the active and neutral leads are identical. An imbalance indicates a fault, and the device cuts the power to that circuit.

Resistance: The opposition to the flow of electrons. Handy in electric kettles, but not in the wiring to a fridge.

Series (connection): Linking by connecting positive to negative, (as in two 12-volt batteries to obtain 24-volts). With series connections, voltage is additive. The current remains the same. See also 'Parallel (connections)'.

Solar Constant: The average amount of solar radiation that reaches the earth's *upper* atmosphere on a surface perpendicular to the sun's rays. It is equal to 1353-watts per square metre. The average maximum at ground level is 800-1000-watts per square metre.

Solar noon: The time at a specific location, where the sun reaches its highest apparent point in the sky.

State of charge: The amount by which a battery is charged (relative to its total capacity).

Surge capacity: The ability to withstand and supply a short-term overload, e.g. of electric current.

Temperature compensation: Adjusting charging current to compensate for temperature change.

Tilt angle: The angle at which a solar array is set to face into the sun relative to the horizontal.

TLC: Three Letter Acronym.

Varistor: A resistor that allows current to pass if the applied voltage is excessive. Varistors are used as lightning protectors in solar cabling. They also allow high voltage static to flow to earth.

Voltage: The pressure that causes current to flow (akin to water pressure in a pipe). Voltage is expressed as volt or V.

Voltage drop: A conductor's resistance to current flow causes a loss of voltage along that conductor. As a result, there are fewer volts at the load end than at the supply end. This difference is called 'voltage drop'.

Watts: The unit of power which is the rate at which work is done or energy transformed to do that work.

Watt-hour (also Wh or W.h): The unit of work or energy equal to one watt for one hour.

References & contacts etc.

Useful sources of information

Australia and New Zealand Solar Energy Society: anzses.org

Carbon offset providers: offsetguide.com.au

Centre for Appropriate Technologies: cfat.org.au

Clean Energy Council: cleanenergycouncil.org.au

Energy ratings of products: energyrating.gov.au

Geothermal Resources Council: geothermal.org

General solar grants information: https://www.cleanenergycoun-cil.org.au/consumers/buying-solar/government-programs?

Recommended books and publications

Solar That Really Works! -3rd Ed. (ISBN: 978-0-6483190-3-0): oriented to caravans, motorhomes and small cabins. *Caravan & Motorhome Electrics* - (originally *Motorhome Electrics*) (ISBN: 978-0-6483190-8-5): this book is written for a general readership, but bought globally by those technically knowledgeable seeking to know about this field. Both are bought by TAFE (in Australia) and auto-electricians worldwide.

The above books are widely available at on-line and physical book-shops in paperback or a variety of eBook formats. Details about where you can find the books are available on our websites – rv-books.com.au and solarbooks.com.au. Both websites have an exten-sive library of exclusive Articles that are regularly updated.

The books are also available in print form from most book shops in Australia and New Zealand. They are also stocked by Jaycar Elec-tronics, Low Energy Developments, Altronics and specialised retail outlets.

ReNew magazine: (quarterly): all aspects of alternative energy: re-new.org.au

Sanctuary magazine (quarterly): sustainable houses and living etc.: renew.org.au/sanctuary-magazine/

Earth Garden, sustainable living and alternatives: rwww.earthgarden.com.au

Solar Living Source Book (ISBN: 0-916571-04-1, Chelsea Green Publishing). American mega-presentation but it has a mass of (sometimes dated) information on who makes what and where.

Relevant standards

AS 4509.1: 2009 Stand-alone power systems: Safety and Installation

AS 4509.2: 2010 Stand-alone power systems: System Design

AS 4086:1993 Secondary batteries for stand-alone power systems

AS/NZS 5033:2012 Installation of stand-alone photovoltaic (PV) arrays

AS/NZS 3000:2018 Electrical Wiring Rules

AS/NZS 3001:2008 - as Amended 2012

AS 1768:2007 Lightning Protection

AS/NZS 1170.2 Wind Loads

AS4777 Grid-connections of Energy Systems via Inverters.

Detailed Table of Contents

Preface	1
Terminology	3
Chapter 1	5
Solar Reality - an overview	5
How much solar is available?	6
Solar - what runs/what doesn't	7
Stand-alone solar systems	9
Grid-connect	10
What solar modules produce	10
Solar access	10
Service importance	11
First all-solar town	11
Chapter 2	13
Lighting	13
Fluorescent tubes and globes	13
Compact fluorescents	14
Light-emitting diodes	14
Light colour	15
Lighting levels	15
Recommended light levels	16
Chapter 3	17
Fridges & freezers	17
Fridge ratings	18
Domestic fridges	19
Chapter 4	21
Air conditioning	21
Air conditioner star ratings	21
Air conditioner types	22
Inverter air conditioners	22
Reducing air conditioner	22

consumption	
Cooling capacity required	23
Reverse-cycle air conditioners	23
Chapter 5	25
Washing machines	25
Washing machine - energy ratings	26
Water efficiency ratings	26
Buying caution	27
Rebates	27
Chapter 6	28
Clothes dryers	28
Dryer types	28
Chapter 7	30
Dishwashers	30
Energy ratings	30
Saving energy	31
Chapter 8	32
Power tools	32
Chapter 9	34
Phantom loads	34
Example phantom loads	34
Remedying phantom loads	35
Chapter 10	37
TVs and computers	37
TV energy draw	37
Computers	37
Chapter 11	39
Water	39
Rainwater tanks	40
Rainwater legislation	41
Pumping water	41
Pump energy draw	42
Pressure systems	42
Constant pressure pumps	43

Variable speed pumps	44	**Chapter 17**	71
Pressure tanks	44	Batteries	71
Irrigation	46	Battery types	71
Valve control	47	Battery capacity	72
Bore pumps	48	Interconnecting batteries	73
Pumping losses	48	Battery location	75
Calculating pump pressure	50	**Chapter 18**	77
		Battery charging	77
Chapter 12	52	The battery charging area	78
Swimming pools	52	**Chapter 19**	79
Keeping water clean	53	Energy monitoring	79
Recent trends	53	**Chapter 20**	81
Chapter 13	55	Generators	81
Ponds	55	Diesel generators for long term use	82
Ultra-violet sterilisers	55		
Pumps - general	56	Battery charging from a generator	82
Solar pumps	57		
Chapter 14	59	Auto-start/stop	83
Solar	59	Generator noise	83
What solar modules produce	59	Generator size	84
		Generators and switch-mode chargers	84
Nominal Operating Cell Temperature	59		
		Chapter 21	85
How much sun?	60	Alternative power - wind & hydro	85
Tracking the sun	60		
Shadowing tolerance and loss	61	Larger wind generators	86
		Micro hydro-electric	86
Weather effects	62	Turbine types	87
Buying solar modules	62	Ram pumps	88
Chapter 15	63	Geothermal space heating	89
Solar modules - voltage & current	63		
		Passive and direct solar heating	90
Increasing voltage	64		
Power Maximisers	65	Energy efficiency first	90
Solar output beware - it's not what it seems	67	Site selection	90
		Chapter 22	92
Chapter 16	68	Alternative energy storage - fuel cells	92
Solar regulation	68		
Maximum Power Point Tracking (MPPT) regulators	69	Fuel cell history	92
		Commercial development	93
		How fuel cells work	94

Chapter 23	95		The inverter decides	121
Inverters	95		Solar array voltage	121
Silicon camels	96		Establishing cable size	121
Stand-alone inverters	96		How most cable is rated	123
Inverters for fixed wiring	97		Auto cable	123
Grid-connect inverters	98		Current ratings	124
Battery backed grid-con-nect	98		Cable (voltage separa-tion)	125
Chapter 24	100		Terminating cables	125
Energy auditing	100		Power posts	126
Existing usage	100		Crimping tool & lugs	127
Chapter 25	104		Protecting cabling	128
Scaling stand-alone sys-tems	104		DC circuit breakers	128
			AC circuit breakers	129
Available irradiation	105		Fuses	129
Needed solar capacity	106		Solar modules need un-obstructed sun	130
Solar regulators	107		Solar mountings	130
Example of solar regula-tor sizing	108		Connecting modules	131
Battery capacity	108		Solar modules lightning protection	132
Battery chargers	109		Testing series-connected solar modules	133
Generators for battery charging	109		Testing parallel-connected modules	133
Fuel cells	110		Earthing solar module frames	134
Chapter 26	111			
Meters & measuring	111		Solar regulator	134
How to measure solar current output	112		Programming solar regu-lators	136
How to measure battery voltage	112		Locating energy monitors	136
			Locating batteries	136
How to measure battery and other heavy currents	113		Example	137
Current shunts	113		Large capacity battery banks	137
Chapter 27	115			
Installing - legal	115		Battery chargers	137
Working on batteries	116		Current shunts	138
Inverter safety	118		Switches for 12/24-volts DC	138
Chapter 28	119			
Installing the system	119		Power outlets for 12/24-volts	139
Reducing voltage drop	119			
All 12/24-volts	120			

Inverters	139
Generators	140
Generators-auto/changeover	141
Water pumps	141
Pressure accumulators	142
Air conditioners	142
Refrigerators	143
Fridge cable size	144
Fusing fridges	145
Chapter 29	147
Constructing a Stand-alone System	147
Chapter 30	150
Grid-connect	150
The grid network	150
Electricity usage	150
Distributed energy storage	151
Grid-connect	151
Going off-grid	152
Climate-optimised solar	153
Strategies to consider now	153
Chapter 31	155
Example systems	155
Comments	156
Example 2 - mid-sized cabin	157
Updates	158
Further update	160
Example 4 - an ongoing (Queensland) system	160
The costs	162
The savings	163
Update	163
Chapter 32	165
Living with solar	165
Seeming lack of solar input	165
Battery maintenance	166
Leaking batteries	166
Solar modules	166
Chapter 33	167
Our solar systems	167
The Broome house's structure	168
Extending the solar	169
Water	171
Irrigation	172
Swimming pool	172
The end of an era	173
Chapter 34	175
Technical Terms Explained	175
References & contacts etc.	181
Useful sources of information	181
Recommended books and publications	181
Relevant standards	182